Springer Theses

Recognizing Outstanding Ph.D. Research

For further volumes:
http://www.springer.com/series/8790

Aims and Scope

The series "Springer Theses" brings together a selection of the very best Ph.D. theses from around the world and across the physical sciences. Nominated and endorsed by two recognized specialists, each published volume has been selected for its scientific excellence and the high impact of its contents for the pertinent field of research. For greater accessibility to non-specialists, the published versions include an extended introduction, as well as a foreword by the student's supervisor explaining the special relevance of the work for the field. As a whole, the series will provide a valuable resource both for newcomers to the research fields described, and for other scientists seeking detailed background information on special questions. Finally, it provides an accredited documentation of the valuable contributions made by today's younger generation of scientists.

Theses are accepted into the series by invited nomination only and must fulfill all of the following criteria

- They must be written in good English.
- The topic should fall within the confines of Chemistry, Physics and related interdisciplinary fields such as Materials, Nanoscience, Chemical Engineering, Complex Systems and Biophysics.
- The work reported in the thesis must represent a significant scientific advance.
- If the thesis includes previously published material, permission to reproduce this must be gained from the respective copyright holder.
- They must have been examined and passed during the 12 months prior to nomination.
- Each thesis should include a foreword by the supervisor outlining the significance of its content.
- The theses should have a clearly defined structure including an introduction accessible to scientists not expert in that particular field.

Haixing Miao

Exploring Macroscopic Quantum Mechanics in Optomechanical Devices

Doctoral Thesis accepted by
School of Physics,
The University of Western Australia

 Springer

Author
Dr. Haixing Miao
Theoretical Astrophysics
Caltech M350-17
E. California Blvd 1200
Pasadena CA 91125
USA

Supervisors
Prof. Dr. David Blair
Australian International Gravitational
  Research Centre (AIGRC)
The University of Western
  Australia (M013)
35 Stirling Highway
Crawley WA 6009
Australia

Prof. Dr. Yanbei Chen
Theoretical Astrophysics
Mail Code 350-17
California Institute of Technology
Pasadena CA 91125-1700
USA

ISSN 2190-5053
ISBN 978-3-642-42645-2
DOI 10.1007/978-3-642-25640-0
Springer Heidelberg Dordrecht London New York

e-ISSN 2190-5061
ISBN 978-3-642-25640-0 (eBook)

© Springer-Verlag Berlin Heidelberg 2012
Softcover reprint of the hardcover 1st edition 2012
This work is subject to copyright. All rights are reserved, whether the whole or part of the material is concerned, specifically the rights of translation, reprinting, reuse of illustrations, recitation, broadcasting, reproduction on microfilm or in any other way, and storage in data banks. Duplication of this publication or parts thereof is permitted only under the provisions of the German Copyright Law of September 9, 1965, in its current version, and permission for use must always be obtained from Springer. Violations are liable to prosecution under the German Copyright Law.
The use of general descriptive names, registered names, trademarks, etc. in this publication does not imply, even in the absence of a specific statement, that such names are exempt from the relevant protective laws and regulations and therefore free for general use.

Printed on acid-free paper

Springer is part of Springer Science+Business Media (www.springer.com)

Decrease your frequency by expanding your horizon. Increase your Q by purifying your mind. Eventually, you will achieve inner peace and view the internal harmony of our world.

—A lesson from a harmonic oscillator

Dedicated to my parents Lanying Zhang and Dehua Miao

Parts of this thesis have been published in the following journal articles:

1. Haixing Miao, Chunong Zhao, Li Ju, Slawek Gras, Pablo Barriga, Zhongyang Zhang, and David G. Blair, *Three-mode optoacoustic parametric interactions with a coupled cavity*, Phys. Rev. A **78**, 063809 (2008).
2. Haixing Miao, Chunnong Zhao, Li Ju and David G. Blair, *Quantum ground-state cooling and tripartite entanglement with three-mode optoacoustic interactions*, Phys. Rev. A **79**, 063801 (2009).
3. Chunnong Zhao, Li Ju, Haixing Miao, Slawomir Gras, Yaohui Fan, and David G. Blair, *Three-Mode Optoacoustic Parametric Amplifier: A Tool for Macroscopic Quantum Experiments*, Phys. Rev. Lett. **102**, 243902 (2009).
4. Farid Ya. Khalili, Haixing Miao, and Yanbei Chen, *Increasing the sensitivity of future gravitational-wave detectors with double squeezed-input*, Phys. Rev. D **80**, 042006 (2009).
5. Haixing Miao, Stefan Danilishin, Thomas Corbitt, and Yanbei Chen, *Standard Quantum Limit for Probing Mechanical Energy Quantization*, Phys. Rev. Lett. **103**, 100402 (2009).
6. Haixing Miao, Stefan Danilishin, Helge Mueller-Ebhardt, Henning Rehbein, Kentaro Somiya, and Yanbei Chen, *Probing macroscopic quantum states with a sub-Heisenberg accuracy*, Phys. Rev. A **81**, 012114 (2010).
7. Haixing Miao, Stefan Danilishin, and Yanbei Chen, *Universal quantum entanglement between an oscillator and continuous fields*, Phys. Rev. A **81**, 052307 (2010).
8. Haixing Miao, Stefan Danilishin, Helge Mueller-Ebhardt, and Yanbei Chen, *Achieving ground state and enhancing optomechanical entanglement by recovering information*, New Journal of Physics, **12**, 083032 (2010).
9. Farid Ya. Khalili, Stefan Danilishin, Haixing Miao, Helge Mueller-Ebhardt, Huan Yang, and Yanbei Chen, *Preparing a Mechanical Oscillator in Non-Gaussian Quantum States*, Phys. Rev. Lett. **105**, 070403 (2010).

Supervisor's Foreword

Quantum mechanics is a successful and elegant theory for describing the behaviors of both microscopic atoms and macroscopic condensed-matter systems. However, there remains the interesting and fundamental question as to how an apparently macroscopic classical world emerges from the microscopic one described by quantum wave functions. Recent achievements in high-precision measurement technologies could eventually lead to answering this question through studies of quantum phenomena in the macroscopic regime.

By coupling coherent light to mechanical degrees of freedom via radiation pressure, several groups around the world have built state-of-the-art optomechanical devices that are very sensitive to the tiny motions of mechanical oscillators. One prominent example is the laser interferometer gravitational-wave detector, which aims to detect weak gravitational waves from astrophysical sources in the universe. With high-power laser beams, and high mechanical quality test masses, future advanced gravitational-wave detectors will achieve extremely high displacement sensitivity—so high that they will be limited by fundamental noise of quantum origin, and the kilogram-scale test masses will have to be considered quantum mechanically. This means, on the one hand, that we should manipulate the optomechanical interaction between the optical field and the test masses coherently at the quantum level, in order to further improve the detector sensitivity; and, on the other hand, that advanced gravitational-wave detectors will be ideal platforms for studying the quantum dynamics of kilogram-scale test masses—truly macroscopic objects.

These two interesting aspects of advanced gravitational-wave detectors, and of more general optomechanical devices, are the main subjects of this dissertation. The author, Dr. Haixing Miao, starts with a quantum model for the optomechanical device, and studies its various quantum features in detail. In the first part of the thesis, different approaches are considered for surpassing the quantum limit on the displacement sensitivity of gravitational-wave detectors; in the second part, experimental protocols are considered for probing the quantum behaviors of macroscopic mechanical oscillators with both linear and non-linear optomechanical interactions. This thesis has inspired much interesting work within the gravitational-wave community, and has been awarded the prestigious Gravitational Wave International Committee (GWIC) thesis prize in 2011. In addition, the formalism developed here may be equally well applied to general quantum limited measurement devices, which are also of interest to the quantum optics community.

Australia, September 2011

Winthrop Professor David Blair
Director, Australian International Gravitational
Research Centre

Preface

Recent significant achievements in fabricating low-loss optical and mechanical elements have aroused intensive interest in optomechanical devices which couple optical fields to mechanical oscillators, e.g., in laser interferometer gravitationalwave (GW) detectors. Not only can such devices be used as sensitive probes for weak forces and tiny displacements, but they also lead to the possibilities of investigating quantum behaviors of macroscopic mechanical oscillators, both of which are the main topics of this thesis. They can shed light on improving the sensitivity of quantum-limited measurement, and on understanding the quantumto-classical transition.

This thesis summarizes and puts into perspective several research projects that I worked on together with the UWA group and the LIGO Macroscopic Quantum Mechanics (MQM) discussion group. In the first part of this thesis, we will discuss different approaches for surpassing the standard quantum limit for the displacement sensitivity of optomechanical devices, mostly in the context of GW detectors. They include: (1) *Modifying the input optics.* We consider filtering two frequency-independent squeezed light beams through a tuned resonant cavity to obtain an appropriate frequency dependence, which can be used to reduce the measurement noise of the GW detector over the entire detection band; (2) *Modifying the output optics.* We study a time-domain variational readout scheme which measures the conserved dynamical quantity of a mechanical oscillator: the mechanical quadrature. This evades the measurement-induced back action and achieves a sensitivity limited only by the shot noise. This scheme is useful for improving the sensitivity of signal-recycled GW detectors, provided the signalrecycling cavity is detuned, and the optical spring effect is strong enough to shift the test-mass pendulum frequency from 1 Hz up to the detection band around 100 Hz; (3) *Modifying the dynamics.* We explore frequency dependence in double optical springs in order to cancel the positive inertia of the test mass, which can significantly enhance the mechanical response and allow us to surpass the SQL over a broad frequency band.

In the second part of this thesis, two essential procedures for an MQM experiment with optomechanical devices are considered: (1) *state preparation*, in which we prepare a mechanical oscillator in specific quantum states. We study the preparations of both Gaussian and non-Gaussian quantum states, and also the creation of quantum entanglements between the mechanical oscillator and the optical field. Specifically, for the Gaussian quantum states, e.g., the quantum ground state, we consider the use of passive cooling and optimal feedback control in cavity-assisted schemes. For non-Gaussian quantum states, we introduce the idea of coherently transferring quantum states from the optical field to the mechanical oscillator. For the quantum entanglement, we consider the entanglement between the mechanical oscillator and the finite degrees-of-freedom cavity modes, and also the infinite degrees-of-freedom continuum optical mode. (2) *state verification*, in which we probe and verify the prepared quantum states. A similar time-dependent homodyne detection method as discussed in the first part is implemented to evade the back action, which allows us to achieve a verification accuracy that is below the Heisenberg limit. The experimental requirements and feasibilities of these two procedures are considered in both small-scale cavity-assisted optomechanical devices, and in large-scale advanced GW detectors.

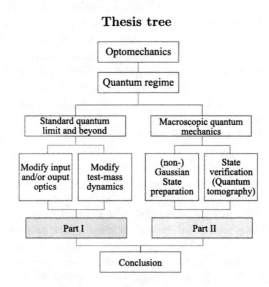

Thesis tree

Acknowledgments

I am very thankful to my supervisors: Chunnong Zhao, David Blair and Ju Li at the University of Western Australia (UWA), and Yanbei Chen at the California Institute of Technology (Caltech). With great patience and enthusiasm, they introduced me to many interesting topics, especially, optomechanical interactions and their classical and quantum theories which make this thesis possible. Whenever I encountered some problems that could not be overcome, their sharp insights and great motivations always lit me up, and helped me to move forward.

I also want to express my thankfulness to Stefan Danilishin, Mihai Bondarescu, Helge Mueller-Ebhardt, Chao Li, Henning Rehbein, Thomas Corbitt, Kentaro Somiya, Farid Khalili, and all the other members in the LIGO-MQM discussion groups. In the two months of visiting the Albert-Einstein Institute (AEI) and MQM telecons, I had intensive discussions with them, which produced many fruitful results in this thesis. I thank especially Stefan who played significant roles in all my work concerning macroscopic quantum mechanics.

I am very thankful to Rana Adhikari, Koji Arai, Kiwamu Izumi, Jenne Driggers, David Yeaton-Massey, Aiden Brook and Steve Vass at Caltech, with whom I spent my enjoyable 4 month experimental investigations of an advanced suspension isolation scheme based upon magnetic levitation. Rana Adhikari and Koji Arai made painstaking efforts in trying to teach me the fundamentals of electronics and feedback control theory.

I would like to thank Antoine Heidmann, Pierre-Franùcois Cohadon, and Chiara Molinelli for their friendly hosting of my visit to the Laboratoire Kastler Brossel, and for helping me to understand how to characterize a mechanical oscillator experimentally.

I thank all my colleagues at UWA: Yaohui Fan, Zhongyang Zhang, Andrew Sunderland, and Andrew Woolley. They are easy-going and friendly, and the friendship with them has made my postgraduate study life colorful and enjoyable.

I would like to thank Ruby Chan for helping to arrange my visits to AEI and Caltech, and also for helping me with many other administrative issues.

I thank André Fletcher (UWA) for helping with proof-reading the original copy of this thesis.

My research has been supported by the Australian Research Council and the Department of Education, Science and Training. Special thanks are due to the Alexander von Humboldt Foundation and the David and Barbara Groce startup fund at Caltech, which has supported my visit to AEI and Caltech.

Finally, I am greatly indebted to my beloved parents and my best friends: Yi Feng, Zheng Cai, Shenniang Xu, Zhixiong Liang, Xingliang Zhu, and Jie Liu, who have been supporting and encouraging me all the way along.

Contents

1 Introduction 1
   References 10

2 **Quantum Theory of Gravitational-Wave Detectors** 13
   2.1 Preface 13
   2.2 Introduction 13
   2.3 An Order-of-Magnitude Estimate 14
   2.4 Basics for Analyzing Quantum Noise 15
        2.4.1 Quantization of the Optical Field
               and the Dynamics 15
        2.4.2 Quantum States of the Optical Field 17
        2.4.3 Dynamics of the Test-Mass 19
        2.4.4 Homodyne Detection 20
   2.5 Examples 20
        2.5.1 Example I: Free Space 21
        2.5.2 Example II: A Tuned Fabry-Pérot Cavity 23
        2.5.3 Example III: A Detuned Fabry-Pérot Cavity 25
   2.6 Quantum Noise in an Advanced GW Detector 26
        2.6.1 Input–Output Relation of a Simple Michelson
               Interferometer 26
        2.6.2 Interferometer With Power-Recycling Mirror
               and Arm Cavities 29
        2.6.3 Interferometer With Signal-Recycling 31
   2.7 Derivation of the SQL: A General Argument 34
   2.8 Beating the SQL by Building Correlations 36
        2.8.1 Signal-Recycling 36
        2.8.2 Squeezed Input 37
        2.8.3 Variational Readout: Back-Action Evasion 38
        2.8.4 Optical Losses 39

	2.9	Optical Spring: Modification of Test-Mass Dynamics	41
		2.9.1 Qualitative Understanding of Optical-Spring Effect	41
	2.10	Continuous State Demolition: Another Viewpoint on the SQL	42
	2.11	Speed Meters	44
		2.11.1 Realization I: Coupled Cavities	44
		2.11.2 Realization II: Zero-Area Sagnac	47
	2.12	Conclusions	48
	References	48
3	**Modifying Input Optics: Double Squeezed-Input**	51
	3.1	Preface	51
	3.2	Introduction	51
	3.3	Quantum Noise Calculation	54
		3.3.1 Filter Cavity	54
		3.3.2 Quantum Noise of the Interferometer	56
	3.4	Numerical Optimizations	58
	3.5	Conclusions	61
	References	61
4	**Modifying Test-Mass Dynamics: Double Optical Spring**	65
	4.1	Preface	65
	4.2	Introduction	65
	4.3	General Considerations	67
	4.4	Further Considerations: Removing the Friction Term	69
	4.5	"Speed-Meter" Type of Response	70
	4.6	Conclusions and Future Work	72
	References	72
5	**Measuring a Conserved Quantity: Variational Quadrature Readout**	75
	5.1	Preface	75
	5.2	Introduction	75
	5.3	Dynamics	77
	5.4	Variational Quadrature Readout	78
	5.5	Stroboscopic Variational Measurement	81
	5.6	Conclusions	82
	References	82
6	**MQM With Three-Mode Optomechanical Interactions**	85
	6.1	Preface	85
	6.2	Introduction	86
	6.3	Quantization of Three-Mode Parametric Interactions	88
	6.4	Quantum Limit for Three-Mode Cooling	90

	6.5	Stationary Tripartite Optomechanical Quantum Entanglement...	95
	6.6	Three-Mode Interactions With a Coupled Cavity	99
	6.7	Conclusions...	104
	References		104
7	**Achieving the Ground State and Enhancing Optomechanical Entanglement**		107
	7.1	Preface	107
	7.2	Introduction..	107
	7.3	Dynamics and Spectral Densities	110
		7.3.1 Dynamics	110
		7.3.2 Spectral Densities..........................	112
	7.4	Unconditional Quantum State and Resolved-Sideband Limit...	114
	7.5	Conditional Quantum State and Wiener Filtering	115
	7.6	Optimal Feedback Control............................	118
	7.7	Conditional Optomechanical Entanglement and Quantum Eraser	119
	7.8	Effects of Imperfections and Thermal Noise	122
	7.9	Conclusions...	123
	References		123
8	**Universal Entanglement Between an Oscillator and Continuous Fields**		127
	8.1	Preface	127
	8.2	Introduction..	127
	8.3	Dynamics and Covariance Matrix......................	129
	8.4	Universal Entanglement..............................	131
	8.5	Entanglement Survival Duration.......................	133
	8.6	Maximally-Entangled Mode...........................	133
	8.7	Numerical Estimates.................................	136
	8.8	Conclusions...	137
	References		137
9	**Nonlinear Optomechanical System for Probing Mechanical Energy Quantization**...................................		141
	9.1	Preface	141
	9.2	Introduction..	141
	9.3	Coupled Cavities	142
	9.4	General Systems....................................	147
	9.5	Conclusions...	148
	References		148

10 State Preparation: Non-Gaussian Quantum State ... 151
- 10.1 Preface ... 151
- 10.2 Introduction ... 151
- 10.3 Order-of-Magnitude Estimate ... 153
- 10.4 General Formalism ... 155
- 10.5 Single-Photon Case ... 157
- 10.6 Conclusions ... 159
- 10.7 Appendix ... 159
  - 10.7.1 Optomechanical Dynamics ... 159
  - 10.7.2 Causal Whitening and Wiener Filter ... 160
  - 10.7.3 State Transfer Fidelity ... 162
- References ... 163

11 Probing Macroscopic Quantum States ... 165
- 11.1 Preface ... 165
- 11.2 Introduction ... 165
- 11.3 Model and Equations of Motion ... 171
- 11.4 Outline of the Experiment With Order-of-Magnitude Estimate ... 174
  - 11.4.1 Timeline of Proposed Experiment ... 174
  - 11.4.2 Order-of-Magnitude Estimate of the Conditional Variance ... 175
  - 11.4.3 Order-of-Magnitude Estimate of State Evolution ... 176
  - 11.4.4 Order-of-Magnitude Estimate of the Verification Accuracy ... 178
- 11.5 The Conditional Quantum State and its Evolution ... 179
  - 11.5.1 The Conditional Quantum State Obtained From Wiener Filtering ... 180
  - 11.5.2 Evolution of the Conditional Quantum State ... 180
- 11.6 State Verification in the Presence of Markovian Noises ... 182
  - 11.6.1 A Time-Dependent Homodyne Detection and Back-Action-Evasion ... 182
  - 11.6.2 Optimal Verification Scheme and Covariance Matrix for the Added Noise: Formal Derivation ... 186
  - 11.6.3 Optimal Verification Scheme With Markovian Noise ... 188
- 11.7 Verification of Macroscopic Quantum Entanglement ... 190
  - 11.7.1 Entanglement Survival Time ... 191
  - 11.7.2 Entanglement Survival as a Test of Gravity Decoherence ... 192
- 11.8 Conclusions ... 193

11.9 Appendix.................................... 193
    11.9.1 Necessity of a Sub-Heisenberg Accuracy for Revealing Non-Classicality................ 193
    11.9.2 Wiener-Hopf Method for Solving Integral Equations........................ 195
    11.9.3 Solving Integral Equations in Section 11.6 198
References 200

12 Conclusions and Future Work 203
12.1 Conclusions................................. 203
12.2 Future Work 205

Chapter 1
Introduction

Measuring weak forces lies in the heart of modern physics: on the small scale, atomic-force microscopy [18] probes microscopic structures, or even Casimir force, by measuring the displacement of a micro-mechanical cantilever [38]; on the large scale, gravitational-wave (GW) detectors search for ripples in spacetime, by measuring the differential displacements of spatially-separated test masses induced by tiny gravitational tidal forces [cf. Fig. 1.1] [24–26]. The core of all these systems is an optomechanical device with mechanical degrees of freedom coupled to a coherent optical field, as shown schematically in Fig. 1.2. With the availability of highly coherent lasers and low-loss optical and mechanical components, optomechanical devices can attain such a high sensitivity that even the quantum dynamics of the macroscopic mechanical oscillator has to be taken into account, which leads to the fundament quantum limit for the measurement sensitivity—the so-called "Standard Quantum Limit".

Standard Quantum Limit(SQL)—The SQL was first realized by Braginsky in the 1960s, when he studied whether quantum mechanics imposes any limit on the force sensitivity of bar-type GW detectors. As we will see, such a limit is directly related to the fundamental Heisenberg uncertainty principle, and it applies universally to all devices that use a mechanical oscillator as a probe mass. Its force noise spectral density $S^F_{\rm SQL}$ reads:

$$S^F_{\rm SQL}(\Omega) = 2\hbar |m[(\Omega^2 - \omega_m^2) + 2i\gamma_m \Omega]|, \tag{1.1}$$

with Ω the angular frequency, m the mass, ω_m the eigenfrequency, and γ_m the damping rate of the mechanical oscillator.

In the case of an interferometric GW detector, such as LIGO [25], the mechanical oscillators are kg-scale test masses suspended with a pendulum frequency around 1 Hz. Since the frequency of the GW signal that we are interested in is around 100 Hz, they can be well approximated as free masses with $\omega_m \sim 0$. In addition, the gravitational tidal force on two test masses separated by L is $F_{\rm tidal} = mL\ddot{h}$ with h the GW strain, which in the frequency domain reads $-mLh\Omega^2$. Therefore, the corresponding h-referred SQL reads:

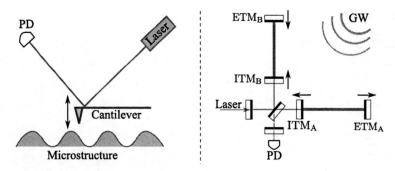

Fig. 1.1 A schematic plot of an atomic-force microscope (*left*), and a gravitational-wave (GW) detector (*right*)

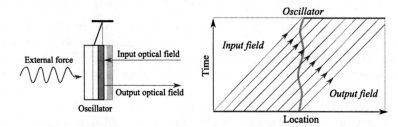

Fig. 1.2 A schematic plot of an optomechanical system (*left*), and the corresponding spacetime diagram (*right*). The output optical field that contains the information of the oscillator motion is measured continuously by a photodetector. For clarity, the input and output optical fields are placed on opposite sides of the oscillator world line

$$S_h^{\text{SQL}}(\Omega) = \frac{2\hbar}{m\Omega^2 L^2}, \qquad (1.2)$$

where we have ignored the damping rate γ_m because the quality factor of a typical suspension is very high.

There are two perspectives on the origin of the SQL. The first is based upon the dynamics of the optomechanical system. At high frequencies, the quantum fluctuation of the optical phase gives rise to phase *shot noise*, which is inversely proportional to the optical power; while at low frequencies, the quantum fluctuation of the optical amplitude creates a random radiation-pressure force on the mechanical oscillator and induces *radiation-pressure noise* which is directly proportional to the optical power. If these two types of noise are not correlated, they will induce a lower bound on the detector sensitivity independent of the optical power. The locus of such a lower bound gives the SQL, as shown schematically in Fig. 1.3. The second perspective is based upon the fact that oscillator positions at different times do not commute with each other—$[\hat{x}(t), \hat{x}(t')] \neq 0 (t \neq t')$. Therefore, according to the Heisenberg uncertainty principle, a precise measurement of the oscillator position at an early time will deteriorate the precision of a later measurement. Since we infer the external force by measuring the changes in the oscillator position, this will impose a limit on the

1 Introduction

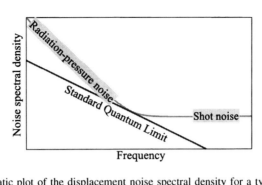

Fig. 1.3 A schematic plot of the displacement noise spectral density for a typical GW detector. When we increase the power, the shot noise will decrease and the radiation-pressure noise will increase, and vise versa. The locus of the power-independent *lower* bound of the total spectrum defines the SQL (*blue*)

force sensitivity. These two perspectives are intimately connected to each other due to the linearity of the system dynamics, as will be shown in Chap. 2.

Surpassing the SQL—From these previous two perspectives on the SQL, we can find different approaches towards surpassing it, as discussed extensively in the literature. The first approach is to modify the input and output optics such that the shot noise and the radiation-pressure noise are correlated, because the SQL exists only when these two noises are uncorrelated. As shown by Kimble et al. [30], by using frequency-dependent squeezed light, the correlation between the shot noise and the radiation-pressure noise allows the sensitivity to be improved by the squeezing factor over the entire detection band. The required frequency dependence can be realized by filtering frequency-independent squeezed light through two detuned Fabry-Pérot cavities before sending into the dark port of the interferometer. Motivated by the work of Corbitt et al. [8], we figure out that such a frequency dependence can also be achieved by filtering two frequency-independent squeezed lights through a tuned Fabry-Pérot cavity. In addition to the detection at the interferometer dark port, another detection at the filter cavity output is essential to maximize the sensitivity. The configuration is shown schematically in Fig. 1.4. An advantage of this scheme is that it only requires a relatively short filter cavity (~ 30 m), in contrast to the km-long filter cavity proposed in Ref. [30]. It can be a feasible add-on to advanced GW detectors. This is discussed in detail in Chap. 3.

The second approach is to modify the dynamics of the mechanical oscillator, e.g., by shifting its eigenfrequency to where the signal is, and amplifying the signal at the shifted frequency. This is particularly useful for GW detectors in which the pendulum frequency of the test masses is very low. If the test-mass frequency is shifted to ω_m, the corresponding SQL surpassing ratio is:

$$\eta \equiv \frac{S^F_{\rm SQL}|_{\rm modified}}{S^F_{\rm SQL}|_{\rm freemass}} = \frac{\Omega^2}{|(\Omega^2 - \omega_m^2) + 2i\gamma_m \Omega|}. \qquad (1.3)$$

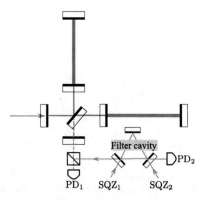

Fig. 1.4 A schematic plot showing the double-squeezed input configuration of an advanced GW detector. Two frequency-independent squeezed (SQZ) *light* are filtered by a tuned Fabry-Pérot cavity before being injected into the *dark* port of the interferometer. Two photodetections (PD) are made, at both the filter cavity, and at the interferometer outputs, to maximize the sensitivity

This is equal to the quality factor $\omega_m/(2\gamma_m)$—which can be approximately 10^7—around the resonant frequency ω_m, thus achieving a significant enhancement. One might naively expect that such a modification of test-mass dynamics can be achieved by a classical feedback control. However, classical control can modify the test-mass dynamics but not increase the sensitivity. This is because a classical control feeds back the measurement noise and signal in the same manner. We have to implement a quantum feedback which modifies the test-mass dynamics without increasing the measurement noise. One possible way to achieve a quantum feedback is to use the optical-spring effect. This happens when a test-mass is coupled to a detuned optical cavity: the intra-cavity power, or equivalently the radiation-pressure force on the test-mass, depends on the location of the test-mass as shown in Fig. 1.5, which creates a spring. One issue with the optical spring is the anti-damping force which destabilizes the system. This arises from the delay in the response with a finite cavity storage time. To stabilize the system, one can use a feedback control method as described in Ref. [4]. An interesting alternative is to implement the idea of a double optical spring by pumping the cavity with two lasers at different frequencies [9, 45]. One laser with a small detuning provides a large positive damping, while another with a large detuning, but with a high power, provides a strong restoring force. The resulting system is self-stabilized with both positive rigidity and positive damping, as shown schematically in the right panel of Fig. 1.5.

One limitation with such a modification of the test-mass dynamics mentioned above is that it only allows a narrow band amplification around the shifted resonant frequency. Recently, as realized by Khalili, this limitation can be overcome by using the frequency dependence of double optical springs, with which the response function of the free test-mass becomes:

$$-m\Omega^2 + K_1(\Omega) + K_2(\Omega) \tag{1.4}$$

1 Introduction

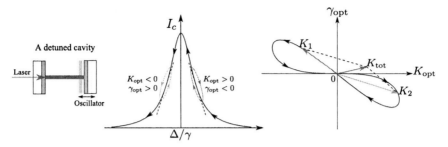

Fig. 1.5 Plot showing the optical spring effect in a detuned optical cavity. The radiation pressure is proportional to the intra-cavity power which depends on the position of the test mass. The non-zero delay in the cavity response gives rise to an (anti-)damping force. By injecting two laser beams at different frequencies, this creates a double optical spring and the system can be stabilized (*right panel*)

with K_1 and K_2 the optical rigidity. Ideally, if $K_1(0) + K_2(0) = 0$, $K_1'(0) + K_2'(0) = 0$ and $K_1''(0) + K_2''(0) = 2m$, the inertia of the test mass is canceled, and a broadband resonance can be achieved. The advantage of this scheme is its immunity to the optical loss compared with modifying the input and/or output optics. Another parameter regime we are interested in is where two lasers with identical power are equally detuned, but with opposite signs. Even though this does not surpass the SQL, yet it allows us to follow the SQL at low frequencies instead of at one particular frequency in the case shown by Fig. 1.3. This is discussed in details in Chap. 4.

A third method is to measure conserved dynamical quantity of the test-mass, also called quantum nondemolition (QND) quantities, which at different times commute with each other. There will be no associated back action, in contrast to the case of measuring non-conserved quantities. For a free mass, the conserved quantity is the momentum (speed), and it can be measured, e.g., by adopting speed-meter configurations [5, 11, 23, 29, 44]. For a high-frequency mechanical oscillator, the conserved quantities are the mechanical quadratures X_1 and X_2, which are defined by the equations:

$$\frac{\hat{x}}{\delta x_q} \equiv \hat{X}_1 \cos \omega_m t + \hat{X}_2 \sin \omega_m t, \quad \frac{\hat{p}}{\delta p_q} \equiv -\hat{X}_1 \sin \omega_m t + \hat{X}_2 \cos \omega_m t, \quad (1.5)$$

with $\delta x_q \equiv \sqrt{\hbar/(2m\omega_m)}$ and $\delta p_q \equiv \sqrt{\hbar m \omega_m/2}$. The quadratures commute with themselves at different times $[\hat{X}_1(t), \hat{X}_1(t')] = [\hat{X}_2(t), \hat{X}_2(t')] = 0$. To measure mechanical quadratures in the cavity-assisted case, one can modulate the optical cavity field strength sinusoidally at the mechanical frequency, as pointed out in the pioneering work of Braginsky [3]. In this case, the measured quantity is proportional to:

$$E(t)\hat{x}(t) = E_0 \hat{x}(t) \cos \omega_m t = E_0 [\hat{X}_1 + \hat{X}_1 \cos 2\omega_m t + \hat{X}_2 \sin 2\omega_m t]/2. \quad (1.6)$$

If the cavity bandwidth is smaller than the mechanical frequency (the so-called good-cavity condition), the $2\omega_m$ terms will have insignificant contributions to the output, and we will measure mostly $\hat{X}_1$, achieving a QND measurement. However, such a good-cavity condition is not always satisfied, especially in broadband GW detectors and small-scale devices. Here, we consider a time-domain variational method for measuring the mechanical quadratures, which does not need such a good-cavity condition. By manipulating the output instead of the input field, the measurement-induced back action can be evaded in the measurement data, achieving essentially the same effect as modulating the input field. This approach is motivated by the work of Vyatchanin et al. [52, 53], in which a time-domain variational method is proposed for detecting GWs with known arrival time.

Macroscopic Quantum Mechanics—We have been discussing the SQL for measuring force with optomechanical devices, and have already seen that the quantum dynamics of the mechanical oscillator plays a significant role. A natural question follows: "Can we use such a device to probe the quantum dynamics of a macroscopic mechanical oscillator, and thereby gain a better understanding of the quantum-to-classical transition, and of quantum mechanics in the macroscopic regime?" The answer would be affirmative if we could overcome a large obstacle in front of us: the thermal decoherence. The coupling between the mechanical oscillator and high-temperature (usually 300 K) heat bath induces random motion which is many order of magnitude higher than that of the quantum zero-point motion.

The solution to such a challenge lies in the optomechanical system itself—that is, the optical field. As the typical optical frequency ω_0 is around 3×10^{14} Hz (infrared), each single quantum $\hbar\omega_0$ has an effective temperature of $\hbar\omega_0/k_B \sim 15,000$ K, which is much higher than the room temperature. This means that the optical field is almost in its ground state, with low entropy, and can create an effectively zero-temperature heat bath at room temperature. This fact illuminates two approaches to preparing a pure quantum ground state of the mechanical oscillator: (i) *Thermodynamical cooling*. In this approach, the mechanical oscillator is coupled to a detuned optical cavity. There is a positive damping force in the optical spring effect when the cavity is red detuned (i.e., laser frequency tuned to be below the resonant frequency of the cavity). If the optomechanical damping $\gamma_{\rm opt}$ is much larger than its original value γ_m, the oscillator is settled down in thermal equilibrium with the zero-temperature optical heat bath, as shown schematically in Fig. 1.6. With this method, many novel experiments have already demonstrated significant reductions of the thermal occupation number of the mechanical oscillator [1, 6, 7, 9, 10, 16, 19, 21, 27, 31, 36, 39, 41, 42, 46–50]. In this thesis, we will discuss such a cooling effect in the three-mode optomechanical interaction where two optical cavity modes are coupled to a mechanical oscillator (i.e., to a mechanical mode) [refer to Chap. 6 for details]. Due to the optimal frequency matching—the frequency gap between two cavity modes is equal to the mechanical frequency—this method significantly enhances the optomechanical coupling, given the same input optical power as the existing two-mode optomechanical interaction used in those cooling experiments. In addition, it is also shown to be less susceptible to classical laser noise. (ii) *Uncertainty reduction based upon information*. Since the optical field is coupled to the oscillator, even if there is no optical spring effect, the information of the oscillator position continuously flows out and is available for detection. From this information, we can reduce our ignorance of the quantum state of the oscillator, and

1 Introduction

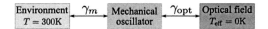

Fig. 1.6 Plot showing that the mechanical oscillator is coupled to both the environmental heat bath with temperature T=300 K, and to the optical field with effective temperature $T_{\text{eff}} = 0$ K. The effective temperature of the mechanical oscillator is given by $T_m = \frac{\gamma_m T + \gamma_{\text{opt}} T_{\text{eff}}}{\gamma_m + \gamma_{\text{opt}}}$. This approaches zero if $\gamma_{\text{opt}} \gg \gamma_m$, which is intuitively expected

map out a classical trajectory of its mean position and momentum in phase space. The remaining uncertainty of the quantum state will be Heisenberg-limited if the measurement is fast and sensitive enough (i.e., the information extraction rate is high), and the thermal noise induces an insignificant contribution to the uncertainty of the quantum state. In this way, the mechanical oscillator is projected to a posterior state, also called *the conditional quantum state*. The usual mathematical treatment of such a process is by using the *stochastic master equation* [13, 14, 17, 23, 37]. Since we are not interested in the transient behavior, the frequency-domain Wiener filter approach provides a neat alternative to obtain the steady-state conditional variance of the oscillator position and momentum (defining the remaining uncertainty). Such an approach also allows us to include non-Markovian noise, which is difficult to deal with by using the stochastic master equation. To localize the quantum state in phase space (zero mean position and momentum), one just needs to feed back the acquired classical information with a classical control. There is a unique optimal controller that makes the residual uncertainty minimum, and close to that of the conditional quantum state [12].

Due to the intimate connection between the quantity of information in a system and its thermodynamical entropy, these two approaches merge together in the case of cavity-assisted cooling scheme. This is motivated by the pioneering work of Marquardt et al. [34] and Wilson-Rae et al. [54]. They showed that there is a quantum limit for the achievable occupation number, which is given by $\gamma^2/(2\omega_m)^2$. In order to achieve the quantum ground state, the cavity bandwidth γ has to be much smaller than ω_m, and this is the so-called good-cavity limit, or resolved-sideband limit. The usual understanding of such a limit is from the thermodynamical point of view, and we point out that it can also be understood as an information loss. By recovering the information at the cavity output, we can achieve a nearly pure quantum state, mostly independent of the cavity bandwidth. This is explained in Chap. 7.

Preparing non-Gaussian quantum states—In the above-mentioned situations, the quantum state is Gaussian. By Gaussian, we mean that its Wigner function, which describes the distribution of the position and momentum in phase space, is a two-dimensional Gaussian function. Since the Wigner function is positive and remains Gaussian, it is describable by a classical probability. A unequivocal signature for 'quantumness' is that the Wigner function can have negative values, e.g. in the well-known 'Schrödinger's Cat' state or the Fock state. To prepare these states, it generally requires nonlinear coupling between the mechanical oscillator and external degrees of freedom. For optomechanical systems, this can be satisfied if the zero-point uncertainty of the oscillator position x_q is the same order of magnitude as the linear

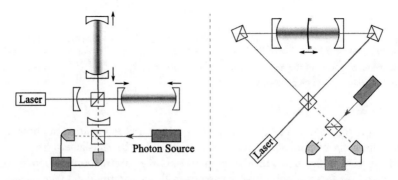

Fig. 1.7 Possible schemes for preparing non-Gaussian quantum states of mechanical oscillators. The *left* panel shows the schematic configuration similar to that of an advanced GW detector with kg-scale suspended test masses in both arm cavities. The *right* panel shows a coupled-cavity scheme proposed in Ref. [50], where a ng-scale membrane is incorporated into a high-finesse cavity. In both cases, a non-Gaussian optical state is injected into the dark port of the interferometer

dynamical range of the optical cavity which is quantified by ratio of the optical wavelength λ to the finesse $\mathcal{F}$:

$$\lambda/(\mathcal{F} x_q) \lesssim 1. \tag{1.7}$$

This condition is also the requirement that the momentum kick induced by a single photon in a cavity be comparable to the zero-point uncertainty of the oscillator momentum. Usually, $\lambda \sim 10^{-6}$ m and $\mathcal{F} \sim 10^6$, which indicates that $x_q \sim 10^{-12}$ m and $m\omega_m \sim 10^{-10}$. This is rather challenging to achieve with the current experimental conditions.

Here we propose a protocol for preparations of a non-Gaussian quantum state which *does not require* nonlinear optomechanical coupling. The idea is to inject a non-Gaussian optical state, e.g., a single-photon pulse created by a cavity QED process [28, 32, 35], into the dark port of the interferometric optomechanical device, as shown schematically in Fig. 1.7. The radiation-pressure force of the single photon on the mechanical oscillator is coherently amplified by the classical pumping from the bright port. As we will show, the qualitative requirement for preparing a non-Gaussian state becomes:

$$\lambda/(\mathcal{F} x_q) \lesssim \sqrt{N_\gamma}. \tag{1.8}$$

Here, $N_\gamma = I_0 \tau/(\hbar\omega_0)$ (I_0 is the pumping laser power, and ω_0 the frequency) is the number of pumping photons within the duration τ of the single-photon pulse, and we gain a significant factor of $\sqrt{N_\gamma}$, as compared with Eq. (1.7), which makes this method experimentally achievable.

Quantum entanglement—As one of the most fascinating features of quantum mechanics, quantum entanglement has triggered many interesting discussions concerning the foundation of quantum mechanics, and it also finds tremendous applications in modern quantum information and computing. If two or more subsystems are entangled, the state of the individual cannot be specified without taking into account the others. Any local measurement on one subsystem will affect others *instantaneously* according to the standard interpretation, which violates the so-called "local realism" rooted in the classical physics. The famous "Einstein-Podolsky-Rosen" (EPR) paradox refers to the quantum entanglement for questioning the completeness of quantum mechanics [15]. To great extents, creating and testing quantum entanglements has been the driving force for gaining better understanding of quantum mechanics.

Interestingly, the optomechanical coupling not only allows us to prepare pure quantum states, but also to create quantum entanglements involving macroscopic mechanical oscillators. Since this directly involves macroscopic degrees of freedom, such entanglements can help us gain insights into the quantum-to-classical transition and various decoherence effects which are significant issues in quantum computing, and many quantum communication protocols [2].

In the case of a cavity-assisted optomechanical system, it is shown that stationary EPR-type quantum entanglement between cavity modes and an oscillator [51], or even between two macroscopic oscillators [22, 33, 40] can be created. We also analyze such optomechanical entanglement in the three-mode system. The optimal frequency matching that enhances the cooling also makes the quantum entanglement easier to achieve experimentally. Additionally, we investigate how the finite cavity bandwidth that induces the cooling limit influences the entanglement in general optomechanical devices. We show that the optomechanical entanglement can be significantly enhanced if we recover the information at the cavity output. In some cases, the existence of the entanglement critically depends on whether we take care of the information loss or not.

Motivated by the work of Ref. [40] which shows that the temperature—the strength of thermal decoherence—only affects the entanglement implicitly, we analyze the entanglement in the simplest optomechanical system with a mechanical oscillator coupled to a coherent optical field. Simple though this system is, analyzing the entanglement is highly nontrivial because the coherent optical field has infinite degrees of freedom. The results are very interesting—the existence of the optomechanical entanglement is indeed not influenced by the temperature directly, and the entanglement exists even when the temperature is high and the mechanical oscillator is highly classical. We obtain an elegant scaling for the entanglement strength, which only depends on the ratios between the characteristic frequencies of the optomechanical interaction and the thermal decoherence. This is discussed in detail in Chap. 8.

State verification—Being able to prepare pure quantum states or entanglements does not tell the full story of an MQM experiment. We need a verification stage, during which the prepared states are probed and verified, to follow up the preparation stage. Suppose the preparation stage finishes at $t=0$, the task of the verifier is to make an ensemble measurement of different mechanical quadratures:

$$\hat{X}_\zeta(0) = \hat{x}(0)\cos\zeta + \frac{\hat{p}(0)}{m\omega_m}\sin\zeta, \qquad (1.9)$$

with $\hat{x}(0)$ and $\hat{p}(0)$ the oscillator position and momentum at $t=0$. By building up the statistics, we can map out their marginal distributions, from which the full Wigner function of the quantum state can be constructed. By comparing the verified quantum state with the prepared one, we can justify the quantum state preparation procedure. This is a rather routine procedure in the quantum tomography of an optical quantum state. However, this is nontrivial with optomechanical devices. Unlike the quantum optics experiments where the optical quadrature can be easily measured with a homodyne detection, in most cases that we are interested in, we only measure the position $\hat{x}(t)$ instead of quadratures and the associated back action will perturb the quantum state that we try to probe. Similar to what is discussed in the first part of this thesis, we also use the time-domain variational measurement to probe the mechanical quadratures with the quantum back action evaded from the measurement data. Given a continuous measurement from $t=0$ to T_{int}, we can construct the following integral estimator:

$$\hat{X} = \int_0^{T_{\text{int}}} dt\, g(t)\hat{x}(t) \propto \hat{x}(0)\cos\zeta' + \frac{\hat{p}(0)}{m\omega_m}\sin\zeta', \tag{1.10}$$

with $\cos\zeta' \equiv \int_0^{T_{\text{int}}} dt\, g(t)\cos\omega_m t$ and $\sin\zeta' \equiv \int_0^{T_{\text{int}}} dt\, g(t)\sin\omega_m t$. In this way, a mechanical quadrature $\hat{X}_{\zeta'}$ can be probed. Here, $g(t)$ is some filtering function, which is determined by the time-dependent homodyne phase and also by the way in which data at different times are combined. By optimizing the filtering function, we can achieve a verification accuracy that is below the Heisenberg limit.

A three-stage MQM experiment—By combining the state preparation and the verification, we can outline a complete procedure for an MQM experiment. In order to probe various decoherence effects, and the quantum dynamics, we can include an evolution stage during which the mechanical oscillator freely evolves. We discuss such a three-stage procedure: the preparation, evolution, and verification in advanced GW detectors. The details are in Chap. 11.

References

1. O. Arcizet, P.-F. Cohadon, T. Briant, M. Pinard, A. Heidmann, Radiation-pressure cooling and optomechanical instability of a micromirror. Nature **444**, 71–74 (2006)
2. D. Bouwmeester, A. Ekert, A. Zeilinger, *The Physics of Quantum Information* (Springer, Berlin, 2002)
3. V.B. Braginsky, Y.I. Vorontsov, K.S. Thorne, Quantum nondemolition measurements. Science **209**, 547–557 (1980)
4. A. Buonanno, Y. Chen, Quantum noise in second generation, signalrecycled laser interferometric gravitational-wave detectors. Phys. Rev. D **64**, 042006 (2001)
5. Y. Chen, Sagnac interferometer as a speed-meter-type, quantumnondemolition gravitational-wave detector. Phys. Rev. D **67**, 122004 (2003)
6. P.F. Cohadon, A. Heidmann, M. Pinard, Cooling of a mirror by radiation pressure. Phys. Rev. Lett. **83**, 3174 (1999)

References

7. LIGO Scientific Collaboration, Observation of a kilogram-scale oscillator near its quantum ground state. New Journal of Physics **11**(7), 073032 (2009)
8. T. Corbitt, N. Mavalvala, S. Whitcomb, Optical cavities as amplitude filters for squeezed fields. Phys. Rev. D **70**, 022002 (2004)
9. T. Corbitt, Y. Chen, E. Innerhofer, H. Mueller-Ebhardt, D. Ottaway, H. Rehbein, D. Sigg, S. Whitcomb, C. Wipf, N. Mavalvala, An All-Optical Trap for a Gram-Scale Mirror. Phys. Rev. Lett. **98**, 150802–4 (2007)
10. T. Corbitt, C. Wipf, T. Bodiya, D. Ottaway, D. Sigg, N. Smith, S. Whitcomb, N. Mavalvala, Optical dilution and feedback cooling of a gram-scale oscillator to 6.9 mK. Phys. Rev. Lett. **99**, 160801–4 (2007)
11. S.L. Danilishin, Sensitivity limitations in optical speed meter topology of gravitational-wave antennas. Phys. Rev. D **69**, 102003 (2004)
12. S. Danilishin, H. Mueller-Ebhardt, H. Rehbein, K. Somiya, R. Schnabel, K. Danzmann, T. Corbitt, C. Wipf, N. Mavalvala, Y. Chen, Creation of a quantum oscillator by classical control, arXiv:0809.2024 (2008)
13. A.C. Doherty, K. Jacobs, Feedback control of quantum systems using continuous state estimation. Phys. Rev. A **60**, 2700 (1999)
14. A.C. Doherty, S.M. Tan, A.S. Parkins, D.F. Walls, State determination in continuous measurement. Phys. Rev. A **60**, 2380 (1999)
15. A. Einstein, B. Podolsky, N. Rosen, Can quantum-mechanical description of physical reality be considered complete?. Phys. Rev. **47**, 777 (1935)
16. I. Favero, C. Metzger, S. Camerer, D. Konig, H. Lorenz, J.P. Kotthaus, K. Karrai, Optical cooling of a micromirror of wavelength size. Appl. Phys. Lett. **90**, 104101–3 (2007)
17. C. Gardiner, P. Zoller, *Quantum Noise* (Springer, Berlin, 2004)
18. F.J. Giessibl, Advances in atomic force microscopy. Rev. Mod. Phys. **75**, 949 (2003)
19. S. Gigan, H.R. Böhm, M. Paternostro, F. Blaser, G. Langer, J.B. Hertzberg, K.C. Schwab, D. Bäuerle, M. Aspelmeyer, A. Zeilinger, Self-cooling of a micromirror by radiation pressure. Nature **444**, 67–70 (2006)
20. S. Gröblacher, S. Gigan, H.R. Bohm, A. Zeilinger, M. Aspelmeyer, Radiation-pressure self-cooling of a micromirror in a cryogenic environment. EPL (Europhys. Lett.) **81**(5), 54003 (2008)
21. S. Gröblacher, J.B. Hertzberg, M.R. Vanner, G.D. Cole, S. Gigan, K.C. Schwab, M. Aspelmeyer, Demonstration of an ultracold microoptomechanical oscillator in a cryogenic cavity. Nat Phys **5**, 485–488 (2009)
22. M.J. Hartmann, M.B. Plenio, Steady state entanglement in the mechanical vibrations of two dielectric membranes. Phys. Rev. Lett. **101**, 200503 (2008)
23. A. Hopkins, K. Jacobs, S. Habib, K. Schwab, Feedback cooling of a nanomechanical resonator. Phys. Rev. B **68**, 235328 (2003) Dec.
24. http://geo600.aei.mpg.de
25. http://www.ligo.caltech.edu
26. http://www.virgo.infn.it
27. G. Jourdan, F. Comin, J. Chevrier, Mechanical mode dependence of bolometric backaction in an atomic force microscopy microlever. Phys. Rev. Lett. **101**, 133904–4 (2008)
28. M. Keller, B. Lange, K. Hayasaka, W. Lange, H. Walther, Continuous generation of single photons with controlled waveform in an ion-trap cavity system. Nature **431**, 1075–1078 (2004)
29. F.Y. Khalili, Y. Levin, Speed meter as a quantum nondemolition measuring device for force. Phys. Rev. D **54**, 004735 (1996)
30. H.J. Kimble, Y. Levin, A.B. Matsko, K.S. Thorne, S.P. Vyatchanin, Conversion of conventional gravitational-wave interferometers into quantum nondemolition interferometers by modifying their input and/or output optics. Phys. Rev. D **65**, 022002 (2001)
31. D. Kleckner, D. Bouwmeester, Sub-kelvin optical cooling of a micromechanical resonator. Nature **444**, 75–78 (2006)

32. C.K. Law, H.J. Kimble, Deterministic generation of a bit-stream of single-photon pulses. J. Mod. Opt. **44**(11), 2067–2074 (1997)
33. S. Mancini, V. Giovannetti, D. Vitali, P. Tombesi, Entangling macroscopic oscillators exploiting radiation pressure. Phys. Rev. Lett. **88**, 120401 (2002)
34. F. Marquardt, J.P. Chen, A.A. Clerk, S.M. Girvin, Quantum theory of cavity-assisted sideband cooling of mechanical motion. Phys. Rev. Lett. **99**, 093902 (2007)
35. J. McKeever, A. Boca, A.D. Boozer, R. Miller, J.R. Buck, A. Kuzmich, H.J. Kimble, Deterministic generation of single photons from one atom trapped in a cavity. Science **303**, 1992–1994 (2004)
36. C.H. Metzger, K. Karrai, Cavity cooling of a microlever. Nature **432**, 1002–1005 (2004)
37. G.J. Milburn, Classical and quantum conditional statistical dynamics. Quantum and Semiclassical Optics: Journal of the European Optical Society Part B **8**(1), 269 (1996)
38. U. Mohideen, A. Roy, Precision measurement of the casimir force from 0.1 to 0.9 mum. Phys. Rev. Lett. **81**, 4549 (1998)
39. C.M. Mow-Lowry, A.J. Mullavey, S. Gossler, M.B. Gray, D.E. McClelland, Cooling of a gram-scale cantilever flexure to 70 mK with a servo-modified optical spring. Phys. Rev. Lett. **100**, 010801–4 (2008)
40. H. Mueller-Ebhardt, H. Rehbein, R. Schnabel, K. Danzmann, Y. Chen, Entanglement of macroscopic test masses and the standard quantum limit in laser interferometry. Phys. Rev. Lett. **100**, 013601 (2008)
41. A. Naik, O. Buu, M.D. LaHaye, A.D. Armour, A.A. Clerk, M.P. Blencowe, K.C. Schwab, Cooling a nanomechanical resonator with quantum back-action. Nature **443**, 193–196 (2006)
42. M. Poggio, C.L. Degen, H.J. Mamin, D. Rugar, Feedback Cooling of a Cantilever's Fundamental Mode below 5 mK. Phys. Rev. Lett. **99**, 017201–4 (2007)
43. P. Purdue, Analysis of a quantum nondemolition speed-meter interferometer. Phys. Rev. D **66**, 022001 (2002)
44. P. Purdue, Y. Chen, Practical speed meter designs for quantum nondemolition gravitational-wave interferometers. Phys. Rev. D **66**, 122004 (2002)
45. H. Rehbein, H. Müller-Ebhardt, K. Somiya, S.L. Danilishin, R. Schnabel, K. Danzmann, Y. Chen, Double optical spring enhancement for gravitational-wave detectors. Phys. Rev. D **78**, 062003 (2008)
46. S.W. Schediwy, C. Zhao, L. Ju, D.G. Blair, P. Willems, Observation of enhanced optical spring damping in a macroscopic mechanical resonator and application for parametric instability control in advanced gravitationalwave detectors. Phys. Rev. A **77**, 013813–5 (2008)
47. A. Schliesser, P. Del'Haye, N. Nooshi, K.J. Vahala, T.J. Kippenberg, Radiation pressure cooling of a micromechanical oscillator using dynamical backaction. Phys. Rev. Lett. **97**, 243905–4 (2006)
48. A. Schliesser, R. Riviere, G. Anetsberger, O. Arcizet, T.J. Kippenberg, Resolved-sideband cooling of a micromechanical oscillator. Nat. Phys. **4**, 415–419 (2008)
49. J.D. Teufel, J.W. Harlow, C.A. Regal, K.W. Lehnert, Dynamical Backaction of Microwave Fields on a Nanomechanical Oscillator. Phys. Rev. Lett. **101**, 197203–4 (2008)
50. J.D. Thompson, B.M. Zwickl, A.M. Jayich, F. Marquardt, S.M. Girvin, J.G.E. Harris, Strong dispersive coupling of a high-finesse cavity to a micromechanical membrane. Nature **452**, 72–75 (2008)
51. D. Vitali, S. Gigan, A. Ferreira, H.R. Bohm, P. Tombesi, A. Guerreiro, V. Vedral, A. Zeilinger, M. Aspelmeyer, Optomechanical entanglement between a movable mirror and a cavity field. Phys. Rev. Lett. **98**, 030405 (2007)
52. S.P. Vyatchanin, A.B. Matsko, Quantum limit on force measurements. JETP **77**, 218 (1993)
53. S.P. Vyatchanin, E.A. Zubova, Quantum variation measurement of a force. Phys. Lett. A **201**, 269–274 (1995)
54. I. Wilson-Rae, N. Nooshi, W. Zwerger, T.J. Kippenberg, Theory of ground state cooling of a mechanical oscillator using dynamical backaction. Phys. Rev. Lett. **99**, 093901 (2007)

Chapter 2
Quantum Theory of Gravitational-Wave Detectors

2.1 Preface

This chapter gives an overview of the quantum theory of gravitational-wave (GW) detectors. It is a modified version of the chapter contributed to a book in progress—*Advanced Gravitational-Wave Detectors*—edited by David Blair. This chapter is written by Yanbei Chen, and myself. It gives a detailed introduction on how to analyze the quantum noise in advanced GW detectors by using input–output formalism, which is also valid for general optomechanical devices. It discusses the origin of the Standard Quantum Limit (SQL) for GW sensitivity from both the dynamics of the optical field, and of the test-mass, which leads us to different approaches for surpassing the SQL: (i) creating correlations between the shot noise and back-action noise; (ii) modifying the dynamics of the test-mass, e.g., through the optical-spring effect; (iii) measuring the conserved dynamical quantity of the test-mass. For each of these approaches, the corresponding feasible configurations to achieve them are discussed in detail. This chapter presents the basic concepts and mathematical tools for understanding later chapters.

2.2 Introduction

The most difficult challenge in building a laser interferometer gravitational-wave (GW) detector is isolating the test masses from the rest of the world (e.g., random kicks from residual gas molecules, seismic activities, acoustic noises, thermal fluctuations, etc.), whilst keeping the device locked around the correct point of operation (e.g., pitch and yaw angles of the mirrors, locations of the beam spots, resonance condition of the cavities, and dark-port condition for the Michelson interferometer). Once all these issues have been solved, we arrive at the issue that we are going to analyze in this chapter: the fundamental noise that arises from quantum fluctuations in the system. A simple estimate, following the steps of Braginsky [2], will already lead us into the quantum world—as it will turn out, the superb sensitivity of GW detectors will be constrained by the back-action noise imposed by the *Heisenberg Uncertainty Principle*, when it is applied to test masses as heavy as 40 kg in the case of Advanced LIGO (AdvLIGO). As Braginsky realized in his analysis, there exists a *Standard Quantum Limit* (SQL) for the sensitivities of GW detectors—further improvements of detector sensitivity beyond this point require us to consider the application of techniques that manipulate the quantum coherence of light to our advantage. In this chapter, we will introduce a set of theoretical tools that will allow us to analyze GW detectors within the framework of quantum mechanics; using these tools, we will describe several examples in which the SQL can be surpassed.

The outline of this chapter is as follows: In Sect. 2.3, we will make an order-of-magnitude estimate of the quantum noise in a typical GW detector, from which we can gain a qualitative understanding of the origin of the SQL; then, in Sect. 2.4, we will introduce the basic concepts and tools to study the quantum dynamics of an interferometer, and the associated quantum noise. In Sect. 2.5, we will analyze the quantum noise in some simple systems to illustrate the procedures for implementing these tools—these simple systems are the fundamental building blocks for an advanced GW detector. We will start to study the quantum noise in a typical advanced GW detector in Sect. 2.6. We will increase the complexity step by step, each of which is connected in sequence to the simple systems analyzed in the previous section. Section 2.7 will present a rigorous derivation of the SQL from a more general context of linear continuous quantum measurements. This can enhance the understanding of the results in the previous section, and also pave the way to different approaches towards surpassing the SQL. In Sect. 2.8, we will talk about the first approach to surpassing the SQL by building correlations among quantum noises, and, in Sect. 2.9, we will illustrate the second approach to beating the SQL by modifying the dynamics of the test mass—in particular, we will discuss the optical spring effect to realize such an approach. Section 2.10 will present an alternative point of view on the origin of the SQL. This will introduce the idea of a speed meter as a third option for surpassing the SQL, in Sect. 2.11—two possible experimental configurations of the speed meter will be discussed. Finally, in Sect. 2.12, we will conclude with a summary of the main results in this chapter.

2.3 An Order-of-Magnitude Estimate

Here, we first make an order-of-magnitude estimate of the quantum limit for the sensitivity. We assume that test-masses have a reduced mass of m, and it is being measured by a laser beam with optical power I_0, and an angular frequency ω_0. Within a measurement duration τ, the number of photons is $N_\gamma = I_0 \tau/(\hbar \omega_0)$. For a coherent light source (e.g., an ideal laser), the number of photons follows a Poisson distribution, and thus its root-mean-square fluctuation is $\sqrt{N_\gamma}$. The corresponding fractional error in the phase measurement, also called the *shot noise*, would be $\delta \phi_{\text{sh}} = 1/\sqrt{N_\gamma}$. For detecting GWs with a period comparable to τ, the displacement

2.3 An Order-of-Magnitude Estimate

noise spectrum of the shot noise is:

$$S_{\text{sh}}^x \approx \frac{\delta \phi_{\text{sh}}^2}{k^2} \tau = \frac{\hbar c^2}{I_0 \omega_0}, \qquad (2.1)$$

with $k \equiv \omega_0/c$ as the wave number.

Meanwhile, the photon-number fluctuation also induces a random radiation-pressure force on the test-mass, which is the *radiation-pressure noise* (also called the back-action noise). Its magnitude is $\delta F_{\text{rp}} = \sqrt{N_\gamma} \hbar k / \tau$, which is equal to the number fluctuation multiplied by the force of a single photon $\hbar k / \tau$. Since the response function of a free mass in the frequency domain is $-1/m\Omega^2$, the corresponding noise spectrum is:

$$S_{\text{rp}}^x \approx \frac{\delta F_{\text{rp}}^2}{m^2 \Omega^4} \tau = \frac{I_0 \omega_0}{c^2} \frac{\hbar}{m^2 \Omega^4}. \qquad (2.2)$$

The total noise spectrum is a sum of S_{sh}^x and S_{rp}^x, namely:

$$S_{\text{tot}}^x = S_{\text{sh}}^x + S_{\text{rp}}^x = \frac{\hbar c^2}{I_0 \omega_0} + \frac{I_0 \omega_0}{c^2} \frac{\hbar}{m^2 \Omega^4} \geq \frac{2\hbar}{m \Omega^2}, \qquad (2.3)$$

as illustrated in Fig. 2.1. The corresponding lower bound that does not depend on the optical power is $S_{\text{SQL}}^x \equiv 2\hbar/(m\Omega^2)$. In terms of GW strain h, it reads

$$S_{\text{SQL}}^h = \frac{1}{L^2} S_{\text{SQL}}^x = \frac{2\hbar}{m \Omega^2 L^2}, \qquad (2.4)$$

with L being the arm length of the interferometer. This introduces us to the SQL [2, 3, 10], which arises as a trade-off between the shot noise and radiation-pressure noise. In the rest of this chapter, we will develop the necessary tools to analyze quantum noise of interferometers from first principles, and to derive the SQL more rigorously. This will allow us to design GW detectors that surpass this limit.

2.4 Basics for Analyzing Quantum Noise

To rigorously analyze the quantum noise in a detector, we need to study its quantum dynamics, of which the basics will be introduced in this section.

2.4.1 Quantization of the Optical Field and the Dynamics

For the optical field, the quantum operator of its quantized electric field is

$$\hat{E} = u(x,y,z) \int_0^{+\infty} \frac{d\omega}{2\pi} \sqrt{\frac{2\pi \hbar \omega}{Ac}} \left[\hat{a}_\omega e^{ikz - i\omega t} + \hat{a}_\omega^\dagger e^{+i\omega t - ikz} \right]. \qquad (2.5)$$

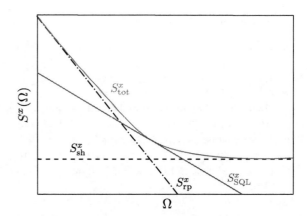

Fig. 2.1 A schematic plot of the displacement noise spectrum for a typical interferometer. Increasing or decreasing the optical power, the power-independent lower bound of the total spectrum will trace over the SQL

Here $\hat{a}_\omega^\dagger$ and $\hat{a}_\omega$ are the creation and annihilation operators, which satisfy $[\hat{a}_\omega, \hat{a}_{\omega'}^\dagger] = 2\pi\,\delta(\omega-\omega')$; $\mathcal{A}$ is the cross-sectional area of the optical beam; $u(x, y, z)$ is the spatial mode, satisfying $(1/\mathcal{A}) \int dx dy |u(x, y, z)|^2 = 1$.

For ground-based GW detectors, the GW signal that we are interested in is in the audio frequency range from 10 to 10^4 Hz. It creates sidebands on top of the carrier frequency of the laser ω_0 (3×10^{14} Hz). Therefore, it is convenient to introduce operators at these sideband frequencies to analyze the quantum noise. The upper and lower sideband operators are $\hat{a}_+ \equiv \hat{a}_{\omega_0+\Omega}$ and $\hat{a}_- \equiv \hat{a}_{\omega_0-\Omega}$, from which we can define the amplitude quadrature $\hat{a}_1$ and phase quadrature $\hat{a}_2$ as:

$$\hat{a}_1 = (\hat{a}_+ + \hat{a}_-^\dagger)/\sqrt{2}, \quad \hat{a}_2 = (\hat{a}_+ - \hat{a}_-^\dagger)/(i\sqrt{2}). \qquad (2.6)$$

They coherently create one photon and annihilate one photon in the upper and lower sidebands, and this is, therefore, also called the two-photon formalism [9]. The electric field can then be rewritten as

$$\hat{E}(x, y, z, t) = u(x, y, z)\sqrt{\frac{4\pi\hbar\omega_0}{\mathcal{A}c}}\left[\hat{a}_1(z, t)\cos\omega_0 t + \hat{a}_2(z, t)\sin\omega_0 t\right].$$

where ω is approximated as ω_0 and the time-domain quadratures are defined as

$$\hat{a}_{1,2}(z, t) \equiv \int_0^{+\infty}\frac{d\Omega}{2\pi}\left(\hat{a}_{1,2}e^{-i\Omega t+ikz} + \hat{a}_{1,2}^\dagger e^{i\Omega t-ikz}\right). \qquad (2.7)$$

These correspond to amplitude and phase modulations in the classical limit.[1]

After having introduced this quantization, we can look further at the dynamics of the optical field. The equations of motion that we will encounter turn out to be very

---

[1] To see such correspondence, suppose the electric field has a large steady-state amplitude A:

$$\hat{E}(z, t) = [A + \hat{a}_1(z, t)]\cos\omega_0 t + \hat{a}_2(z, t)\sin\omega_0 t \approx A\left[1 + \frac{\hat{a}_1(z, t)}{A}\right]\cos\left[\omega_0 t - \frac{\hat{a}_2(z, t)}{A}\right].$$

2.4 Basics for Analyzing Quantum Noise

Fig. 2.2 Two basic dynamical processes of the optical field in analyzing the quantum noise of an interferometer

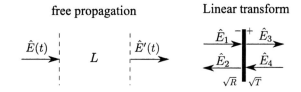

simple, and only two are relevant, as shown in Fig. 2.2: (i) *A free propagation*. Given a free propagation distance of L, the new field $\hat{E}'(t)$ is

$$\hat{E}'(t) = \hat{E}(t - \tau), \quad (2.8)$$

with $\tau \equiv L/c$; (ii) *Continuity condition on the mirror surface*.

$$\hat{E}_2(t) = \sqrt{T}\hat{E}_4(t) - \sqrt{R}\hat{E}_1(t), \quad (2.9)$$

$$\hat{E}_3(t) = \sqrt{R}\hat{E}_4(t) + \sqrt{T}\hat{E}_1(t), \quad (2.10)$$

with transmissivity T, reflectivity R, and a sign of convention as indicated in the figure. These equations relate the optical field before and after the mirror. Due to the linearity of this system, they are both identical to the classical equations of motion.

In later discussions, different quantities of the optical field will always be compared at the same location, and they will all share the same spatial mode. In addition, the propagation phase shift can be absorbed into the time delay. Therefore, we will ignore the factors $u(x, y, z)\sqrt{\frac{4\pi \hbar \omega_0}{Ac}}$, and $e^{\pm i k z}$, hereafter.

2.4.2 Quantum States of the Optical Field

To determine the expectation value and the quantum fluctuation of the Heisenberg operators (related to the quantum noise), e.g., $\langle \psi | \hat{O} | \psi \rangle$, not only should we specify the evolution of $\hat{O}$, but we also need to specify the quantum state $|\psi\rangle$. Of particular interest to us are vacuum, coherent, and squeezed states.

Vacuum state—The vacuum state $|0\rangle$ is, by definition, the state with no excitation and for every frequency, $\hat{a}_\Omega |0\rangle = 0$. The associated fluctuation is:

$$\langle 0 | \hat{a}_i(\Omega) \hat{a}_j^\dagger(\Omega') | 0 \rangle_{\text{sym}} = \pi \delta_{ij} \delta(\Omega - \Omega'), \quad (i, j = 1, 2). \quad (2.11)$$

Equivalently, the double-sided spectral densities [2] for $\hat{a}_{1,2}$ are

---

[2] For any pair of operators $\hat{O}_1$ and $\hat{O}_2$, the double-sided spectral density is defined through

$$\frac{1}{2\pi}\langle 0 | \hat{O}_1(\Omega') \hat{O}_2^\dagger(\Omega) | 0 \rangle_{\text{sym}} \equiv \frac{1}{2\pi}\langle 0 | \hat{O}_1(\Omega') \hat{O}_2^\dagger(\Omega) + \hat{O}_2^\dagger(\Omega) \hat{O}_1(\Omega') | 0 \rangle \equiv \frac{1}{2} S_{O_1 O_2}(\Omega) \delta(\Omega - \Omega').$$

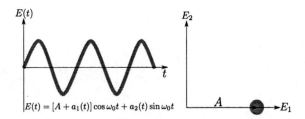

Fig. 2.3 A schematic plot of the electric field and the fluctuations of amplitude and phase quadrature (*shaded area*). The *left* panel shows the time evolution of E, and the *right* panel shows E in the space expanded by the amplitude and phase quadratures (E_1, E_2)

$$S_{a_1}(\Omega) = S_{a_2}(\Omega) = 1, \quad S_{a_1 a_2}(\Omega) = 0. \tag{2.12}$$

Coherent state—The coherent state is defined by [1] as:

$$|\alpha\rangle \equiv \hat{D}[\alpha]|0\rangle \equiv \exp\left[\int \frac{d\Omega}{2\pi}(\alpha_\Omega \hat{a}_\Omega^\dagger - \alpha_\Omega^* \hat{a}_\Omega)\right]|0\rangle, \tag{2.13}$$

which satisfies $\hat{a}_{\Omega'}|\alpha\rangle = \alpha(\Omega')|\alpha\rangle$. The operator $\hat{D}$ is unitary, so $\hat{D}^\dagger \hat{D} = \hat{I}$. We can use this to make a unitary transformation for studying the problem

$$|\psi\rangle \to \hat{D}^\dagger |\psi\rangle, \quad \hat{O} \to \hat{D}^\dagger \hat{O} \hat{D}, \tag{2.14}$$

which leaves the physics invariant. This means that the coherent state can be replaced by the vacuum state, as long as we perform corresponding transformations of $\hat{O}$ into $\hat{D}^\dagger \hat{O} \hat{D}$. For the annihilation and creation operators, we have $\hat{D}^\dagger(\alpha)\hat{a}_\Omega \hat{D}(\alpha) = \hat{a}_\Omega + \alpha_\Omega$ and $\hat{D}^\dagger(\alpha)\hat{a}_\Omega^\dagger \hat{D}(\alpha) = \hat{a}_\Omega^\dagger + \alpha_\Omega^*$, i.e., the original operators plus some complex constants.

An ideal single-mode laser with a central frequency ω_0 can be modeled as a coherent state, and $\alpha_\Omega = \pi \bar{a} \delta(\Omega - \omega_0)$, with $\bar{a} = \sqrt{2 I_0/(\hbar \omega_0)}$ and I_0 is the optical power. Under transformation $\hat{D}$, the electric field reads [cf. Eq. (2.7)]:

$$\hat{E}(t) = [\bar{a} + \hat{a}_1(t)] \cos \omega_0 t + \hat{a}_2(t) \sin \omega_0 t, \tag{2.15}$$

which is simply a sum of a classical amplitude and quantum quadrature fields. This is what we intuitively expect for the optical field from a single-mode laser, namely "quantum fluctuations" superimposed onto a "classical carrier". In Fig. 2.3, we show $E(t)$ and the associated fluctuations in the amplitude and phase quadratures schematically. As we will see later, these fluctuations are attributable to the quantum noise and the associated SQL.

Squeezed state—A more complicated state would be the squeezed state:

$$|[\chi]\rangle \equiv \exp\left[\int_0^{+\infty} \frac{d\Omega}{2\pi}(\chi_\Omega \hat{a}_+^\dagger \hat{a}_-^\dagger - \chi_\Omega^* \hat{a}_+ \hat{a}_-)\right]|0\rangle \equiv \hat{S}[\chi]|0\rangle. \tag{2.16}$$

2.4 Basics for Analyzing Quantum Noise

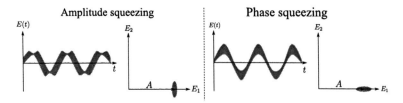

Fig. 2.4 The fluctuations of the amplitude and phase quadratures (*shaded areas*) of the squeezed state. The *left* two panels show the case of amplitude squeezing; the *right* two panels show the phase squeezing

Similar to the coherent-state case, we can also better understand a squeezed state by making a unitary transformation of the basis through $\hat{S}$. By redefining $\chi_\Omega \equiv \xi_\Omega e^{-2i\phi_\Omega}$ ($\xi_\Omega, \phi_\Omega \in \mathfrak{R}$), for quadratures, this leads to:

$$\hat{S}^\dagger \hat{a}_1 \hat{S} = \hat{a}_1(\cosh\xi + \sinh\xi \cos 2\phi) - \hat{a}_2 \sinh\xi \sin 2\phi, \quad (2.17)$$

$$\hat{S}^\dagger \hat{a}_2 \hat{S} = \hat{a}_2(\cosh\xi - \sinh\xi \cos 2\phi) - \hat{a}_1 \sinh\xi \sin 2\phi. \quad (2.18)$$

Let us look at two special cases: (1) $\phi = \pi/2$. We have

$$\hat{S}^\dagger \hat{a}_1 \hat{S} = e^{-\xi} \hat{a}_1, \quad \hat{S}^\dagger \hat{a}_2 \hat{S} = e^{\xi} \hat{a}_2, \quad (2.19)$$

in which the amplitude quadrature fluctuation is squeezed by $e^{-\xi}$ while the phase quadrature is magnified by e^ξ; (2) $\phi = 0$. The situation will just be the opposite. Both cases are shown schematically in Fig. 2.4.

2.4.3 Dynamics of the Test-Mass

Similarly, due to the linear dynamics, the quantum equations of motion for the test masses (relative motion) are formally identical to their classical counterparts:

$$\dot{\hat{x}}(t) = \hat{p}(t)/m, \quad \dot{\hat{p}}(t) = \hat{I}(t)/c + mL\ddot{h}(t). \quad (2.20)$$

Here $\hat{x}$ and $\hat{p}$ are the position and momentum operators, which satisfy $[\hat{x}, \hat{p}] = i\hbar$; $\hat{I}(t)/c$ is the radiation pressure, which is a linear function of the optical quadrature fluctuations; $mL\ddot{h}(t)$ is the GW tidal force. Since the detection frequency (~ 100 Hz) is much larger than the pendulum frequency (~ 1 Hz) of the test-masses in a typical detector, they are treated as free masses.

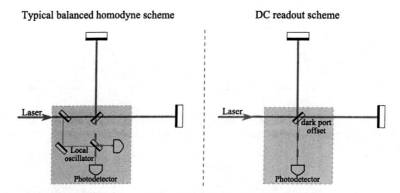

Fig. 2.5 A schematic plot of two homodyne readout schemes

2.4.4 Homodyne Detection

In this section, we will consider how to detect the phase shift of the output optical field which contains the GW signal. To make a phase sensitive measurement, we need to measure the quadratures of the optical field, instead of its power. This can be achieved by a *homodyne detection* in which the output signal light is mixed with a *local oscillator*, thus producing a photon flux that depends linearly on the phase (i.e., on the GW strain). Specifically, for a local oscillator $L(t) = L_1 \cos\omega_0 t + L_2 \sin\omega_0 t$ and output $\hat{b}(t) = \hat{b}_1(t)\cos\omega_0 t + \hat{b}_2(t)\sin\omega_0 t$, the photocurrent is $i(t) \propto |L(t) + \hat{b}(t)|^2 = 2L_1\hat{b}_1(t) + 2L_2\hat{b}_2(t) + \cdots$. The rest of the terms, represented by "$\cdots$", contain either frequency components that are strictly DC and around $2\omega_0$, and terms quadratic in $\hat{b}$. In such a way, we can measure a given quadrature $\hat{b}_\zeta(t) = \hat{b}_1(t)\cos\zeta + \hat{b}_2(t)\sin\zeta$, by choosing the correct local oscillator, such that $\tan\zeta = L_2/L_1$.

In order to realize the above ideal superposition, there are two possible schemes: introducing the local oscillator from the injected laser (external scheme as shown in the right panel of Fig. 2.5;) or intentionally offsetting the two arms at the very beginning, with a very small phase mismatch, which results in the so-called DC readout scheme, as shown in the left panel of Fig. 2.5.

2.5 Examples

Before analyzing the quantum noise in an advanced interferometric GW detector, it is illustrative to first consider three examples: (1) A test mass coupled to an optical field in free space; (2) A tuned Fabry-Pérot cavity with a movable end mirror as the test mass; (3) A detuned Fabry-Pérot cavity with a movable end mirror. These three examples summarize the main physical processes in an advanced GW detector. Understanding them will not only help us to get familiar with the tools for analyzing

2.5 Examples

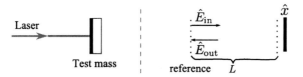

Fig. 2.6 A schematic plot of the interaction between the test mass and a coherent optical field in free space (*left*); and the associated physical quantities (*right*)

quantum noise in a GW detector, but can also provide intuitive pictures which will be useful in understanding more complicated configurations.

2.5.1 Example I: Free Space

The model is shown schematically in Fig. 2.6. The laser-pumped input optical field can be written as [cf. Eq. (2.15)]:

$$\hat{E}_{\text{in}}(t) = \left[\sqrt{2I_0/(\hbar\omega_0)} + \hat{a}_1(t)\right]\cos\omega_0 t + \hat{a}_2(t)\sin\omega_0 t. \tag{2.21}$$

The output field $\hat{E}_{\text{out}}(t)$ is simply:

$$\hat{E}_{\text{out}}(t) = \hat{E}_{\text{in}}(t - 2\tau - 2\hat{x}/c), \tag{2.22}$$

with a delay time $\tau \equiv L/c$. We define output quadratures $\hat{b}_1$ and $\hat{b}_2$ through:

$$\hat{E}_{\text{out}}(t) = \left[\sqrt{2I_0/(\hbar\omega_0)} + \hat{b}_1(t)\right]\cos\omega_0 t + \hat{b}_2(t)\sin\omega_0 t. \tag{2.23}$$

Since the displacement of the test mass is small, and the uncertainty of $\omega_0\hat{x}/c$ is much smaller than unity, we can make a Taylor expansion of Eq. (2.22) in a series of $\omega_0\hat{x}/c$. Up to the leading order, we obtain the following input–output relations:

$$\hat{b}_1(t) = \hat{a}_1(t - 2\tau), \tag{2.24}$$

$$\hat{b}_2(t) = \hat{a}_2(t - 2\tau) - 2\sqrt{\frac{2I_0}{\hbar\omega_0}}\frac{\omega_0}{c}\hat{x}(t - \tau), \tag{2.25}$$

where, for simplicity, we have assumed that $\omega_0 L/c = n\pi$, with n an integer.

The equation of motion for the test-mass displacement $\hat{x}$ is simply:

$$m\ddot{\hat{x}}(t) = \hat{F}_{\text{rp}}(t) + \frac{1}{2}mL\ddot{h}(t). \tag{2.26}$$

Here we have chosen an inertial reference frame, as indicated in Fig. 2.6, such that the gravitational tidal force is equal to $\frac{1}{2}mL\ddot{h}(t)$; the radiation-pressure force $\hat{F}_{\rm rp}$ on the test-mass is given by:

$$\hat{F}_{\rm rp}(t) = 2\frac{\mathcal{A}}{4\pi}|\hat{E}_{\rm in}(t-\tau)|^2 = 2\frac{I_0}{c}\left[1 + \sqrt{\frac{2\hbar\omega_0}{I_0}}\hat{a}_1(t-\tau)\right], \qquad (2.27)$$

where in the second equality we have kept to the first order of the amplitude quadrature. There is a DC component in the radiation-pressure force, which can be balanced in the experiment (e.g., by the wire tension in the case of a suspended pendulum). We are interested in the perturbed part, proportional to the amplitude quadrature, which accounts for the radiation-pressure noise.

We can solve Eqs. (2.24), (2.25) and (2.26) by transforming them into the frequency domain, after which we obtain:

$$\vec{b}(\Omega) = \mathbf{M}\,\vec{a}(\Omega) + \vec{D}\,h(\Omega), \qquad (2.28)$$

where $\vec{a} = (\hat{a}_1, \hat{a}_2)^{\mathbf{T}}$, $\vec{b} = (\hat{b}_1, \hat{b}_2)^{\mathbf{T}}$ (superscript $^{\mathbf{T}}$ denoting transpose); the transfer matrix $\mathbf{M}$ and transfer vector $\vec{D}$ can be read off from the following explicit expression of Eq. (2.28):

$$\begin{bmatrix}\hat{b}_1(\Omega)\\\hat{b}_2(\Omega)\end{bmatrix} = e^{2i\Omega\tau}\begin{bmatrix}1 & 0\\-\kappa & 1\end{bmatrix}\begin{bmatrix}\hat{a}_1(\Omega)\\\hat{a}_2(\Omega)\end{bmatrix} + \begin{bmatrix}0\\e^{i\Omega\tau}\sqrt{2\kappa}\end{bmatrix}\frac{h(\Omega)}{h_{\rm SQL}}, \qquad (2.29)$$

with

$$\kappa = \frac{8I_0\omega_0}{mc^2\Omega^2}, \quad h_{\rm SQL} = \sqrt{\frac{8\hbar}{m\Omega^2 L^2}}. \qquad (2.30)$$

As we can see, the GW signal is contained in the output phase quadrature $\hat{b}_2$. It can be decomposed into signal and noise components:

$$\hat{b}_2(\Omega) = \langle \hat{b}_2(\Omega)\rangle + \Delta\hat{b}_2(\Omega), \qquad (2.31)$$

where $\langle \hat{b}_2(\Omega)\rangle$ is the expectation value of the output, which is proportional to the GW signal h, and Δb_2 is the quantum fluctuation with zero expectation. By defining $\langle \hat{b}_2(\Omega)\rangle \equiv \mathcal{T}h$, we introduce the following quantity:

$$\mathcal{T} = e^{i\Omega\tau}\sqrt{2\kappa}\,\frac{1}{h_{\rm SQL}}, \qquad (2.32)$$

which is the transfer function from the GW strain h to the output phase quadrature. This particular form indicates that the output phase modulation is proportional to the GW strain, delayed by a constant time τ. The noise part

$$\Delta\hat{b}_2(\Omega) = e^{2i\Omega\tau}\hat{a}_2(\Omega) - e^{2i\Omega\tau}\kappa\,\hat{a}_1(\Omega), \qquad (2.33)$$

2.5 Examples

contains two parts: (i) the first one is the *shot noise* $\hat{n}_{\text{sh}} \equiv e^{2i\Omega\tau}\hat{a}_2$, which arises from the phase-quadrature fluctuation of the input optical field and has a flat spectrum [cf. Eq. (2.12)]:

$$S_{\text{sh}}(\Omega) = 1; \tag{2.34}$$

and (ii) the second one is the *radiation-pressure noise* $\hat{n}_{\text{rp}} \equiv (-e^{2i\Omega\tau}\kappa\,\hat{a}_1)$. This arises from the amplitude-quadrature fluctuation, and has the following noise spectrum:

$$S_{\text{rp}}(\Omega) = \kappa^2, \tag{2.35}$$

with a frequency dependence of $1/\Omega^4$.

Given the coherent state of the input optical field, the amplitude and phase-quadrature fluctuations are not correlated. Therefore, we can obtain the total noise spectrum simply by summing up S_{sh} and S_{rp}. By normalizing with the transfer function $\mathcal{T}$, the signal-referred noise spectrum can be written as:

$$S^h(\Omega) = \frac{1}{|\mathcal{T}|^2} S_{\Delta\hat{b}_2}(\Omega) = \left[\frac{1}{\kappa} + \kappa\right]\frac{h_{\text{SQL}}^2}{2} \geq h_{\text{SQL}}^2. \tag{2.36}$$

The shot-noise contribution (first term) is inversely proportional to the optical power ($\kappa \propto I_0$) and the radiation-pressure noise (second term) is proportional to I_0. The balance between them gives the SQL for detecting GWs with this simple model. We will find that although this model is simple, it summarizes the main features of a GW detector.

2.5.2 Example II: A Tuned Fabry-Pérot Cavity

Now we consider the case of a tuned Fabry-Pérot cavity. In Fig. 2.7, we show the model schematically. In comparison with the previous case, an additional mirror with transmissivity (of power) T, and reflectivity R, are placed in front of the test-mass, in effect "wrapping" around the original system. We define the new input and output optical fields $\hat{E}'_{\text{in,out}}$, in a similar way to that of $\hat{E}_{\text{in,out}}$, by simply replacing $\vec{a}, \vec{b}$ with new amplitude and phase quadratures $\vec{\alpha}, \vec{\beta}$. We need to determine a new input–output relation between $\hat{\alpha}_{1,2}$ and $\hat{\beta}_{1,2}$. From the continuity condition on the front mirror surface (cf. Eqs. (2.9) and (2.10)), we have:

$$\hat{E}_{\text{in}} = \sqrt{R}\,\hat{E}_{\text{out}} + \sqrt{T}\,\hat{E}'_{\text{in}}, \tag{2.37}$$

$$\hat{E}'_{\text{out}} = \sqrt{T}\,\hat{E}_{\text{out}} - \sqrt{R}\,\hat{E}'_{\text{in}}. \tag{2.38}$$

Fig. 2.7 A schematic plot of the tuned-cavity model (*left*); and the associated physical quantities (*right*)

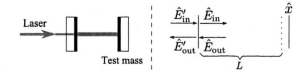

Correspondingly, $\vec{a}, \vec{b}$ are related to new quadrature fields $\vec{\alpha}, \vec{\beta}$ by:

$$\vec{a} = \sqrt{R}\,\vec{b} + \sqrt{T}\,\vec{\alpha}, \tag{2.39}$$

$$\vec{\beta} = \sqrt{T}\,\vec{b} - \sqrt{R}\,\vec{\alpha}. \tag{2.40}$$

Together with Eq. (2.28), it would be straightforward to obtain the new input–output relation. Generally, the expression is rather cumbersome. We will focus on the case in which the transmissivity T is small (i.e., a high-finesse cavity). In addition, since the GW sideband frequency Ω we are interested in is around 100 Hz, $\Omega\tau$ is much smaller than unity even when the cavity length L is 4 km. Therefore, we can make a Taylor expansion of the new input–output relation as a series of the dimensionless quantities T and $\Omega\tau$. Up to the leading order, this leads to:

$$\begin{bmatrix}\hat{\beta}_1(\Omega)\\\hat{\beta}_2(\Omega)\end{bmatrix} = e^{2i\phi}\begin{bmatrix}1 & 0\\-\mathcal{K} & 1\end{bmatrix}\begin{bmatrix}\hat{\alpha}_1(\Omega)\\\hat{\alpha}_2(\Omega)\end{bmatrix} + e^{-i\phi}\begin{bmatrix}0\\\sqrt{2\mathcal{K}}\end{bmatrix}\frac{h}{h_{\rm SQL}}. \tag{2.41}$$

We have introduced:

$$\phi \equiv \arctan(\Omega/\gamma),\quad \mathcal{K} \equiv \frac{2\gamma\,\iota_c}{\Omega^2(\Omega^2+\gamma^2)}, \tag{2.42}$$

with the cavity bandwidth $\gamma \equiv Tc/(4L)$, parameter $\iota_c \equiv 8\omega_0 I_c/(mLc)$, and intra-cavity power $I_c \equiv 4I_0/T$.

The same as the previous free-space case, we need to read out the phase quadrature of the output field which contains the GW signal. The corresponding *signal-referred* noise spectrum S^h has a similar form to the previous free-space case, but with κ replaced by $\mathcal{K}$ [cf. Eq. (2.36)], i.e.:

$$S^h(\Omega) = \left[\frac{1}{\mathcal{K}} + \mathcal{K}\right]\frac{h_{\rm SQL}^2}{2} \geq h_{\rm SQL}^2. \tag{2.43}$$

For frequencies around $\Omega \sim \gamma$, the shot noise spectrum almost decreases by a factor of $1/T^2$ in comparison with the free-space case. This is attributable to the coherent amplification of the optical power and the signal. The additional mirror serves as a quantum feedback, which allows signals to build up coherently, whilst noise adds up incoherently over time (On the other hand, a classical feedback will not normally increase the signal-to-noise ratio, as feeding back what is already known will not increase knowledge.). For frequencies $\Omega > \gamma$, the shot noise increases as Ω^2, rather than remaining constant in the previous case. This is due to the non-zero response time of the cavity, and signal with frequencies higher than γ are averaged out. Therefore, the cavity bandwidth roughly determines the detection bandwidth.

2.5 Examples

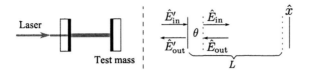

Fig. 2.8 A schematic plot of a detuned cavity; and the associated physical quantities (*right*). Here, θ is the detuned phase

2.5.3 Example III: A Detuned Fabry-Pérot Cavity

If the cavity is not tuned, as shown schematically in Fig. 2.8, namely, $\omega_0 \tau = \theta + n\pi$ ($\theta \neq 0$) with n an integer, the free propagation will not only induce a phase shift, but also a rotation of the quadratures. This simply arises from the following fact: given a free-space propagation of τ, and from the relation $\hat{E}^{\text{out}}(t) = \hat{E}^{\text{in}}(t - \tau)$, the quadrature evolves as:

$$\begin{bmatrix} \hat{b}_1(\Omega) \\ \hat{b}_2(\Omega) \end{bmatrix} = e^{i\Omega\tau} \begin{bmatrix} \cos\omega_0\tau & -\sin\omega_0\tau \\ \sin\omega_0\tau & \cos\omega_0\tau \end{bmatrix} \begin{bmatrix} \hat{a}_1(\Omega) \\ \hat{a}_2(\Omega) \end{bmatrix}, \tag{2.44}$$

which is a delay and rotation.

Correspondingly, Eqs. (2.39) and (2.40) are modified:

$$\vec{a} = \sqrt{R}\,\mathbf{R}_{2\theta}\,\vec{b} + \sqrt{T}\,\mathbf{R}_\theta\,\vec{\alpha}, \tag{2.45}$$

$$\vec{\beta} = \sqrt{T}\,\mathbf{R}_\theta\,\vec{b} - \sqrt{R}\,\vec{\alpha}, \tag{2.46}$$

where $\mathbf{R}_\theta$ is the rotation matrix, defined as

$$\mathbf{R}_\theta \equiv \begin{pmatrix} \cos\theta & -\sin\theta \\ \sin\theta & \cos\theta \end{pmatrix}. \tag{2.47}$$

Similarly, if the detuned phase is small, with $\theta \ll 1$, we can make a Taylor expansion of these equations in series of θ, T and $\Omega\tau$. After some manipulation, the new input–output relation can be expressed in the following compact form:

$$\vec{\beta}(\Omega) = \frac{1}{C}[\mathbf{M}\,\vec{\alpha}(\Omega) + \vec{D}\,h(\Omega)], \tag{2.48}$$

where

$$C = \Omega^2[(\Omega + i\gamma)^2 - \Delta^2] + \Delta\iota_c, \tag{2.49}$$

$$\mathbf{M} = \begin{bmatrix} -\Omega^2(\Omega^2 + \gamma^2 - \Delta^2) - \Delta\iota_c & 2\gamma\Delta\Omega^2 \\ -2\gamma\Delta\Omega^2 + 2\gamma\iota_c & -\Omega^2(\Omega^2 + \gamma^2 - \Delta^2) - \Delta\iota_c \end{bmatrix}, \tag{2.50}$$

$$\vec{D} = \begin{bmatrix} \Delta\Omega \\ (-\gamma + i\Omega)\Omega \end{bmatrix} \frac{2\sqrt{\gamma\iota_c}}{h_{\text{SQL}}}, \tag{2.51}$$

with detuning frequency $\Delta \equiv \theta/\tau$. Here we have ignored the tiny frequency-dependent phase correction $\Omega\theta/\omega_0$. Unlike in the previous two cases, the GW signal here appears in both amplitude and phase quadratures. To readout the GW signal, we can make a homodyne detection of a certain output quadrature:

$$\hat{B}_\zeta(\Omega) = \hat{B}_1(\Omega)\cos\zeta + \hat{B}_2(\Omega)\sin\zeta. \tag{2.52}$$

Given a coherent state input, the corresponding signal-referred noise spectrum density is

$$S^h(\Omega) = \frac{(\cos\zeta,\ \sin\zeta)\mathbf{MM}^{\mathbf{T}}(\cos\zeta,\ \sin\zeta)^{\mathbf{T}}}{|D_1\cos\zeta + D_2\sin\zeta|^2}, \tag{2.53}$$

with $D_{1,2}$ being the components of the vector $\vec{D}$. This expression recovers the previous two cases: (1) the tuned cavity, by setting $\Delta = 0$, and phase quadrature measurement $\zeta = 0$; (2) the free-space case, by setting the cavity bandwidth $\gamma \to \infty$. We will postpone discussing the physical significance of this formula until we consider a signal-recycled GW detector, which can actually be mapped into a detuned Fabry-Pérot cavity.

2.6 Quantum Noise in an Advanced GW Detector

After having introduced some basic principles and examples, we are now ready to analyze the quantum noise of a typical advanced GW detector: a Michelson interferometer with Fabry-Pérot arm cavities, a power-recycling mirror (PRM), and a signal-recycling mirror (SRM). This is shown schematically in Fig. 2.9.

To make a direct one-to-one correspondence between the input–output relation of an advanced GW detector and the three examples we have considered, we will gradually introduce important optical elements, and discuss them in the following sequence: (1) a simple Michelson interferometer with only end test-masses (Sect. 2.6.1); (2) a power-recycled interferometer with both power-recycling mirror and arm cavities (Sect. 2.6.2); (3) a power- and signal-recycled interferometer (Sect. 2.6.3).

2.6.1 Input–Output Relation of a Simple Michelson Interferometer

A simple Michelson interferometer is shown schematically in Fig. 2.10. Ideally, the interferometer is set up to have identical arms, so that at the zero working point of the

2.6 Quantum Noise in an Advanced GW Detector

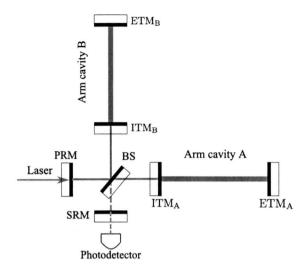

Fig. 2.9 A schematic plot of an advanced GW detector. The beam splitter (*BS*) splits the laser light into two beams. The internal test-mass (*ITM*) and end test-mass (*ETM*) with optical coatings on their surface form the Fabry-Pérot arm cavities which amplifies both the signal and optical power. The power recycling mirror (*PRM*) can further increase the circulating power. The signal-recycling mirror (*SRM*) folds the signal back into the interferometer, and it significantly enriches the dynamics of the system, as discussed in the main text

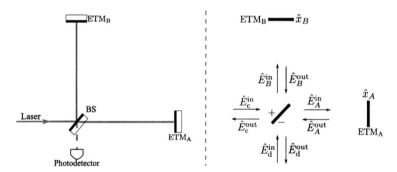

Fig. 2.10 A schematic plot of a simple Michelson interferometer (*left*); and its mathematical model with propagating optical fields (*right*)

interferometer (i.e., when locked on a dark fringe), fields entering from each port will only return to that port. The carrier light enters and exits from the *common (bright) port*, while the differential port remains *dark*. The differential motion of the test mass $\hat{x}_A - \hat{x}_B$, which contains the GW signal causes a differential phase modulation, and therefore induces an output signal out of the differential (dark) port, at which we make homodyne detections.

We follow steps similar to those in Sect. 2.5.1 to derive the input–output relation here. As we will see, the input–output relation of the differential displacement we are interested in is exactly the same as the free-space scenario considered in Sect. 2.5.1. The laser-pumped input optical field into the common port is:

$$\hat{E}_c^{in}(t) = [\sqrt{2I_0/(\hbar\omega_0)} + \hat{c}_1(t)]\cos\omega_0 t + \hat{c}_2(t)\sin\omega_0 t. \quad (2.54)$$

With no laser pumping, the input field into the differential port is simply:

$$\hat{E}_d^{in}(t) = [\hat{a}_1(t)\cos\omega_0 t + \hat{a}_2(t)\sin\omega_0 t]. \quad (2.55)$$

The fields, after passing through the half-half beam splitter, and while propagating towards ETM_A and ETM_B, are:

$$\hat{E}_{A,B}^{in}(t) = \frac{\hat{E}_c^{in}(t) \mp \hat{E}_d^{in}(t)}{\sqrt{2}}. \quad (2.56)$$

The fields returning from the ETM are

$$\hat{E}_{A,B}^{out}(t) = \hat{E}_{A,B}^{in}(t - 2\tau - 2\hat{x}_{A,B}/c), \quad (2.57)$$

where $\tau \equiv L/c$ is the time for light to propagate from the beam splitter to each of the ETMs. To the leading order in $\hat{x}_{A,B}$, we have:

$$\hat{E}_d^{out}(t) = \frac{\hat{E}_B^{out}(t) - \hat{E}_A^{out}(t)}{\sqrt{2}} \equiv [\hat{b}_1(t)\cos\omega_0 t + \hat{b}_2(t)\sin\omega_0 t], \quad (2.58)$$

with

$$\hat{b}_1(t) = \hat{a}_1(t - 2\tau), \quad (2.59)$$

$$\hat{b}_2(t) = \hat{a}_2(t - 2\tau) - \sqrt{\frac{2I_0}{\hbar\omega_0}}\frac{\omega_0}{c}\hat{x}_d(t - \tau), \quad (2.60)$$

where we have assumed $\omega_0 L/c = n\pi$, with n an integer; and have defined the differential-mode motion:

$$\hat{x}_d(t) \equiv \hat{x}_B(t) - \hat{x}_A(t). \quad (2.61)$$

Radiation-pressure forces acting on the two test-masses have both common and differential components, which are proportional to $\hat{c}_1$ and $\hat{a}_1$ respectively. If test masses have nearly the same mass, m, then $\hat{c}_1$ ($\hat{a}_1$) will only induce common-mode (differential-mode) motion. Mathematically, we have, up to leading order in fluctuations/modulations:

$$\hat{F}_{A,B}(t) = 2\frac{I_0}{c}\left[1 + \sqrt{\frac{\hbar\omega_0}{I_0}}\frac{\hat{c}_1(t-\tau) \mp \hat{a}_1(t-\tau)}{\sqrt{2}}\right]. \quad (2.62)$$

2.6 Quantum Noise in an Advanced GW Detector

For the differential mode:

$$\hat{F}_B(t) - \hat{F}_A(t) = 2\frac{\sqrt{2\hbar\omega_0 I_0}}{c}\hat{a}_1(t-\tau). \quad (2.63)$$

This means that the motion of the differential mode under both the radiation-pressure force and the tidal force, $F^h_{A,B} = \mp mL\ddot{h}(t)/2$, from GW is:

$$m\ddot{\hat{x}}_d(t) = \hat{F}_B(t) - \hat{F}_A(t) + F^h_B(t) - F^h_A(t) = 2\frac{\sqrt{2\hbar\omega_0 I_0}}{c}\hat{a}_1(t-\tau) + mL\ddot{h}(t). \quad (2.64)$$

Equations (2.59), (2.60) and (2.61) are identical to Eqs. (2.24)–(2.26), if we identify the previous $2\hat{x}$ by the differential displacement $\hat{x}_d$ here. Since the GW signal also increases by a factor of 2, due to the differential motion of two arms, the signal strength will not change. Therefore, the signal-referred noise spectrum obtained in the free-space case also applies [cf. Eq. (2.36)], namely:

$$S^h(\Omega) = \left[\frac{1}{\kappa} + \kappa\right]\frac{h^2_{\rm SQL}}{2}, \quad (2.65)$$

except for the fact that here:

$$\kappa = \frac{4I_0\omega_0}{mc^2\Omega^2}, \quad h_{\rm SQL} = \sqrt{\frac{4}{m\Omega^2 L^2}}. \quad (2.66)$$

2.6.2 Interferometer With Power-Recycling Mirror and Arm Cavities

In order to decrease the shot noise, we need to increase the optical power. It would be difficult to achieve a high optical power by solely increasing the input power. Instead, we can add a power-recycling mirror, as first proposed by [13] (see Fig. 2.11). The output optical field from the common port gets coherently reflected back into the interferometer, and amplifies the circulation power. Since we are only concerned with the optical field in the differential port, the effect of the PRM can easily be included by simply replacing the input I_0 in $\hat{E}^c_{\rm in}(t)$ (Eq. 2.54) by:

$$I'_0 \equiv \frac{4}{T_{\rm PRM}}I_0, \quad (2.67)$$

where $T_{\rm PRM}$ is the power transmissivity of the PRM. Further improvement of the sensitivity comes from the arm cavities formed by the ITMs and ETMs. These cavities are tuned on resonance, further increasing the optical power circulating in the arms, and also coherently amplifying the GW signal by increasing the effective arm length.

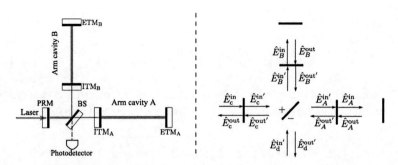

Fig. 2.11 A schematic plot of a Michelson interferometer with a power-recycled mirror (*PRM*) and additional ITMs to form arm cavities (*left*). The corresponding propagating fields are indicated in the *right* diagram

The input–output relation at the differential port also has *the same form as the tuned Fabry-Pérot cavity* discussed in Sect. 2.5.2. This can be shown as follows: The new optical fields $\hat{E}_A^{\text{in}'}$ and $\hat{E}_A^{\text{out}'}$ are related to $\hat{E}_A^{\text{in}}$ and $\hat{E}_A^{\text{out}}$ simply by:

$$\hat{E}_A^{\text{in}} = \sqrt{R_I}\,\hat{E}_A^{\text{out}} + \sqrt{T_I}\,\hat{E}_A^{\text{in}'}, \tag{2.68}$$

$$\hat{E}_A^{\text{out}'} = \sqrt{T_I}\,\hat{E}_A^{\text{out}} - \sqrt{R_I}\,\hat{E}_A^{\text{in}'}, \tag{2.69}$$

where $\sqrt{R_I}$ and $\sqrt{T_I}$ are the reflectivity and transmissivity of the ITM, respectively. Similar relations hold for the fields propagating in the arm cavity B. These new fields are connected to the new input $\hat{E}_c^{\text{in}'}$ and $\hat{E}_d^{\text{in}'}$ by:

$$\hat{E}_{A,B}^{\text{in}'} = \frac{\hat{E}_c^{\text{in}'}(t) \mp \hat{E}_d^{\text{in}'}(t)}{\sqrt{2}}. \tag{2.70}$$

In addition, the new output in the differential port is:

$$\hat{E}_d^{\text{out}'} = \frac{\hat{E}_B^{\text{out}'}(t) - \hat{E}_A^{\text{out}'}(t)}{\sqrt{2}}. \tag{2.71}$$

The relations between the new inputs $\hat{E}_d^{\text{in}'}$, $\hat{E}_d^{\text{out}'}$ and outputs $\hat{E}_d^{\text{in}}$, $\hat{E}_d^{\text{out}}$, at the differential port, are simply defined by:

$$\hat{E}_d^{\text{in}} = \sqrt{R_I}\,\hat{E}_d^{\text{out}} + \sqrt{T_I}\,\hat{E}_d^{\text{in}'}, \tag{2.72}$$

$$\hat{E}_d^{\text{out}'} = \sqrt{T_I}\,\hat{E}_d^{\text{out}} - \sqrt{R_I}\,\hat{E}_d^{\text{in}'}. \tag{2.73}$$

These have the same form as Eqs. (2.37) and (2.38). Therefore, as long as we are only concerned with the fields at the differential port, the new input and output are related to the previous ones without arm cavities in a similar way as that of a single

2.6 Quantum Noise in an Advanced GW Detector

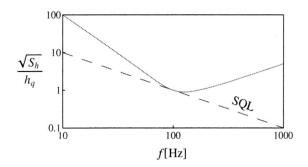

Fig. 2.12 The GW strain-referred sensitivity of an advanced GW detector with power-recycled mirror and arm cavities, given the specifications detailed in the main text. Here, we have normalized the spectrum by h_q, which is defined to be h_{SQL} at 100 Hz

tuned Fabry-Pérot cavity. There is only one difference: in the single tuned cavity analysis, we assumed the front mirror is fixed, while in the GW detector, both ITMs and ETMs can move and the relative motion is detected, which has a reduced mass of $m/2$ in each arm. By further taking into account a factor of two increase in the sensitivity from two arms, the resulting signal-referred noise spectrum reads [cf. Eq. (2.43)]:

$$S^h(\Omega) = \left[\frac{1}{\mathcal{K}} + \mathcal{K}\right] \frac{h_{SQL}^2}{2}, \qquad (2.74)$$

with

$$\mathcal{K} \equiv \frac{2\gamma \iota_c}{\Omega^2(\Omega^2 + \gamma^2)}, \quad h_{SQL} = \sqrt{\frac{8\hbar}{m\Omega^2 L^2}}. \qquad (2.75)$$

The cavity bandwidth, $\gamma \equiv T_I c/(4L)$, and the parameter ι_c are the same as in Eq. (2.42), but with $I_c \equiv 8I_0/(T_{PRM}T_I)$, which is enhanced by both the power-recycling and arm cavities. To illustrate this sensitivity, we can choose the following specifications for different parameters (close to those of the AdvLIGO): mass of individual test mass $m = 40$ kg, intra-cavity optical power $I_c = 800$ kW, arm cavity length $L = 4$ km, optical angular frequency $\omega_0 = 1.9 \times 10^{15} s^{-1}$ (wavelength equal to 1μm), arm cavity bandwidth $\gamma/(2\pi) = 100$ Hz. The corresponding sensitivity is shown in Fig. 2.12, and it achieves the SQL round 100 Hz.

2.6.3 Interferometer With Signal-Recycling

Now we are ready to analyze the quantum noise in an advanced GW detector with both power- and signal-recycling. The schematic plot of the configuration is shown in Fig. 2.13, where a signal-recycling mirror is added onto the differential port, as first proposed by [17, 23]. After previously introducing all the techniques, the effect on the detector sensitivity also becomes apparent. With the same idea, we can

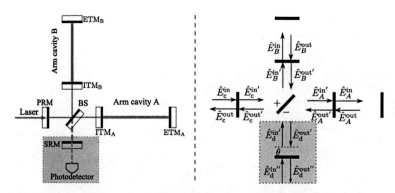

Fig. 2.13 A schematic plot of a Michelson interferometer with a power- (*PRM*) and signal-recycled mirrors (*SRM*) and additional ITMs to form arm cavities (*left*). The corresponding propagating fields are indicated in the *right*-hand diagram, with θ being the detuned phase

relate the new input and output fields $\hat{E}_d^{\text{in}''}$, $\hat{E}_d^{\text{out}''}$ to the $\hat{E}_d^{\text{in}'}$, $\hat{E}_d^{\text{out}'}$ fields analyzed in the previous section. The relation reads:

$$\hat{E}_d^{\text{in}'}(t) = \sqrt{R_S}\,\hat{E}_d^{\text{out}'}(t - 2\theta/\omega_0) + \sqrt{T_S}\,\hat{E}_d^{\text{in}''}(t - \theta/\omega_0), \tag{2.76}$$

$$\hat{E}_d^{\text{out}''}(t) = \sqrt{T_S}\,\hat{E}_d^{\text{out}'}(t - \theta/\omega_0) - \sqrt{R_S}\,\hat{E}_d^{\text{in}''}(t), \tag{2.77}$$

with θ the detuned phase and R_S and T_S the reflectivity and transmissivity of the SRM, respectively. The corresponding quadratures will undergo rotations identical to those shown in the single detuned cavity analysis [cf. Eqs. (2.45) and (2.46)]. The expression for this input–output relation is very lengthy and complicated [6, 7]. We will follow the approach in Ref. [11] and map the entire signal-recycled interferometer into a single detuned cavity, which will then allow us to directly use the results obtained in Sect. 2.5.3.

The idea behind this mapping is based upon the fact that the length, L_{SR}, of the signal-recycling cavity formed by the ITMs and the SRM is, of the order of 10 m, which is very short compared with kilometer long arm cavity. The propagation phase shift of the sidebands around 100 Hz, $e^{i\Omega L_{\text{SR}}}$, is negligible [18, 19]. As shown schematically in Fig. 2.14, the signal-recycling cavity can be replaced by one mirror with a set of effective reflectivities and transmissivities, which are related to $T_{S,I}$ and $R_{S,I}$ by:

$$r_{\text{eff}} = -\frac{\sqrt{R_S} + \sqrt{R_I}\,e^{2i\theta}}{1 + \sqrt{R_I}\,e^{2i\theta}}, \quad t'_{\text{eff}} = \frac{\sqrt{R_I} + \sqrt{R_S R_I}\,e^{2i\theta}}{1 + \sqrt{R_S R_I}\,e^{2i\theta}}, \tag{2.78}$$

$$t_{\text{eff}} = t'_{\text{eff}} = \frac{\sqrt{T_S T_I}\,e^{i\theta}}{1 + \sqrt{R_S R_I}\,e^{2i\theta}}. \tag{2.79}$$

From the resonant condition of the effective cavity:

2.6 Quantum Noise in an Advanced GW Detector

Fig. 2.14 Mapping from a three-mirror cavity to a two-mirror cavity by defining the effective transmissivities (t_{eff}, t'_{eff}) and reflectivities (r_{eff}, r'_{eff}) of the signal-recycling cavity

$$r'_{\text{eff}} e^{2i\Omega_{\text{res}} L/c} = 1, \quad (2.80)$$

we can define its effective bandwidth γ_{eff} and detuning Δ_{eff} through:

$$\Omega_{\text{res}} \equiv -\Delta_{\text{eff}} - i\,\gamma_{\text{eff}}. \quad (2.81)$$

Up to the leading order of T_I, Δ_{eff} and γ_{eff} can be expressed in terms of $\sqrt{T_{S,I}}$, $\sqrt{R_{S,I}}$ and θ through

$$\Delta_{\text{eff}} = \frac{2\sqrt{R_S}\gamma_I \sin 2\theta}{1 + R_S + 2\sqrt{R_S}\cos 2\theta}, \quad \gamma_{\text{eff}} = \frac{(1 - R_S)\gamma_I}{1 + R_S + 2\sqrt{R_S}\cos 2\theta}, \quad (2.82)$$

with the bandwidth of the arm cavity $\gamma_I \equiv c\,T_I/(4L)$.

After this equivalent mapping into a single detuned cavity, it is straightforward to obtain the input–output relation for the signal-recycling interferometer, through the following replacement in Eq. (2.48):

$$\gamma \to \gamma_{\text{eff}}, \quad \Delta \to \Delta_{\text{eff}}. \quad (2.83)$$

For illustration, we assume an ideal phase-quadrature detection with $\zeta = \pi/2$; the corresponding GW strain-referred sensitivity is given by [cf. Eq. (2.53)]:

$$S^h(\Omega) = \frac{4\gamma_{\text{eff}}^2(\iota_c + \Delta_{\text{eff}}\Omega^2)^2 + [\Delta_{\text{eff}}\iota_c + \Omega^2(\gamma_{\text{eff}}^2 - \Delta_{\text{eff}}^2 + \Omega^2)]^2}{4\gamma_{\text{eff}}\iota_c\Omega^2(\gamma_{\text{eff}}^2 + \Omega^2)} h_{\text{SQL}}^2. \quad (2.84)$$

One interesting example is when $\theta = \pi/2$, giving

$$\gamma_{\text{eff}} = \frac{1 + \sqrt{R_S}}{1 - \sqrt{R_S}}\gamma, \quad \Delta_{\text{eff}} = 0. \quad (2.85)$$

The resulting noise spectrum is the same as the previous case without signal-recycling, but with an increased detection bandwidth [a factor of $(1 + \sqrt{R_S})/(1 - \sqrt{R_S})$]. This is the so-called *Resonant-Sideband-Extraction* (RSE) scheme which will be implemented in AdvLIGO. The spectrum is shown in Fig. 2.15. In general, $\theta \neq \pi/2$ and we also show the noise spectrum in the case for a SR cavity detuned phase $\theta = 1.1$. As we can see, there are two minima in the sensitivity curve. Interestingly, they surpass the SQL around the most sensitive frequency (~ 100 Hz for

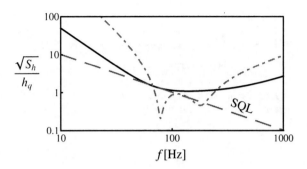

Fig. 2.15 The GW strain-referred sensitivity of a signal-recycling GW detector given a phase-quadrature detection $\zeta = \pi/2$. The *solid* curve corresponds to the RSE scheme with $\theta = \pi/2$ and $\sqrt{R_S} = 0.6$. The *dashed* curve shows the case with $\theta = 1.1$ and $\sqrt{R_S} = 0.9$

given specifications). To gain a better insight into these features which surpasses the SQL, we need to examine the origin of the SQL from a more general picture of linear continuous measurements. Not only will it allow us to understand this particular example, but it can also provide an insight into the SQL, and thus help us to find other approaches to surpass it.

2.7 Derivation of the SQL: A General Argument

Throughout the previous discussions, we have learnt that there are two types of noise, namely shot noise and radiation-pressure noise. They together give rise to the SQL of the detector sensitivity. Actually, the SQL exists in general linear continuous measurements, and it is directly related to the fundamental *Heisenberg Uncertainty Principle* [3]. We will elaborate on this point in this section.

The model of a typical measurement process is shown schematically in Fig. 2.16. The signal—a classical force G (e.g., from the GW) is driving the probe (e.g., the test mass) which is in turn coupled to an external detector (e.g., the optical field). The detector reads out the probe motion by monitoring its displacement $\hat{x}$, and at the same time it exerts a back-action force $\hat{F}$ onto the conjugate momentum of the probe. The signal force G can then be extracted from the detector output $\hat{Y}$, which contains both the signal and the fundamental measurement noise $\hat{Z}$ (i.e., the shot noise). Mathematically, the displacement-referred output $\hat{Y}$, i.e., the measurement result, can be written as:

$$\hat{Y}(t) = \hat{x}_0 + \hat{Z}(t) + \int_{-\infty}^{t} dt' R_{xx}(t-t')[\hat{F}(t') + G(t')]. \quad (2.86)$$

Here $\hat{x}_0$ is the free-evolution value of the probe displacement when the detector is detached, and it has the following two-time commutator:

2.7 Derivation of the SQL: A General Argument

Fig. 2.16 A schematic model of a linear continuous measurement process

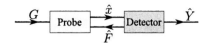

$$[\hat{x}_0(t), \hat{x}_0(t')] = -i\hbar R_{xx}(t-t'), \tag{2.87}$$

where $R_{xx}(t)$ is the response function of the probe to the external force. Here, we require that the detector is tunable, and has a parameter ϵ, with:

$$\hat{Z}_\epsilon(t) = \frac{\epsilon}{\epsilon_0}\hat{Z}_{\epsilon_0}(t), \quad \hat{F}_\epsilon(t) = \frac{\epsilon_0}{\epsilon}\hat{F}_{\epsilon_0}(t). \tag{2.88}$$

The fact that $\hat{Y}(t)$ is the measurement result itself dictates that

$$[\hat{Y}(t), \hat{Y}(t')] = 0, \tag{2.89}$$

because measuring $\hat{Y}(t)$ continuously should not impose any additional noise [3]. In addition, since $\hat{Z}$ and $\hat{F}$ are operators that belong to a different system (the detector) from the probe, they both commute with the probe displacement $\hat{x}_0$. It is therefore required that the two-time commutator of $\hat{Z}$ and $\hat{F}$ terms in $\hat{Y}$ must cancel the commutator from $\hat{x}_0$. In fact, because $\hat{Z}$ and $\hat{F}$ have different scalings in ϵ, and because the cancelation must happen at all orders of ϵ, we can obtain:

$$[\hat{Z}(t), \hat{Z}(t')] = [\hat{F}(t), \hat{F}(t')] = 0, \quad [\hat{Z}(t), \hat{F}(t')] = i\hbar\delta(t-t'). \tag{2.90}$$

This indicates that the shot noise and the back-action noise do not commute at the same moment. In the frequency domain, this can be written as:

$$[\hat{Z}(\Omega), \hat{Z}(\Omega')] = [\hat{F}(\Omega), \hat{F}(\Omega')] = 0, \quad [\hat{Z}(\Omega), \hat{F}(\Omega')] = 2\pi i\hbar\delta(\Omega-\Omega'). \tag{2.91}$$

If we introduce the single-sided (cross) spectral densities $S_{ZZ}(\Omega)$, $S_{FF}(\Omega)$ and $S_{ZF}(\Omega)$, the above commutation relations dictate the Heisenberg Uncertainty Principle:

$$S_{ZZ}(\Omega)S_{FF}(\Omega) - |S_{ZF}(\Omega)|^2 \geq \hbar^2. \tag{2.92}$$

This generally will not set a bound on the noise spectrum in measuring the probe displacement [cf. Eq. (2.86)] which is given by:

$$S^x(\Omega) = S_{ZZ}(\Omega) + 2\Re[R_{xx}(\Omega)S_{ZF}(\Omega)] + |R_{xx}(\Omega)|^2 S_{FF}(\Omega) \tag{2.93}$$

with $R_{xx}(\Omega)$ the Fourier transform of $R_{xx}(t)$. However, when there is no correlation between the shot noise and the back-action noise, namely $S_{ZF} = 0$, it induces the SQL for the displacement measurement:

$$S^x(\Omega) \geq 2|R_{xx}(\Omega)|\sqrt{S_{ZZ}(\Omega)S_{FF}(\Omega)} \geq 2\hbar|R_{xx}(\Omega)| \equiv S^x_{\text{SQL}}. \tag{2.94}$$

In the case of GW detection, the external force is the GW tidal force on the test masses and we have $G(t) = mL\ddot{h}(t)$ (m is the reduced mass). Therefore, the SQL for the GW signal-referred sensitivity reads

$$S_{SQL}^h = \frac{2\hbar}{m^2 \Omega^4 L^2 |R_{xx}(\Omega)|}. \tag{2.95}$$

For a free mass, $R_{xx}(\Omega) = -1/(m\Omega^2)$; this gives the free-mass SQL:

$$S_{SQL}^h = \frac{2\hbar}{m\Omega^2 L^2}, \tag{2.96}$$

which justifies the order-of-magnitude estimate we obtained in Eq. (2.4).

From the above derivation, we immediately realize that there are two possible approaches to surpassing the free-mass SQL: (1) Creating correlations between the shot noise $\hat{Z}$ and the back-action noise $\hat{F}$ with non-zero S_{ZF}. Correspondingly, the inequality in Eq. (2.94) is not satisfied, and the total noise spectrum will not be bounded by the SQL; (2) Modifying the dynamics of the probe: The free-mass SQL will no longer be relevant if the probe has a different response to the external force than the free mass. In particular, for an oscillator with a resonance frequency ω_m and decay rate γ_m, the response function reads $R_{xx}(\Omega) = 1/[-m(\Omega^2 - \omega_m^2 + i\gamma_m \Omega)]$, and then around resonance frequency:

$$\frac{S_{SQL}^h|_{\text{oscillator}}}{S_{SQL}^h|_{\text{free mass}}} = \frac{\sqrt{(\Omega^2 - \omega_m^2)^2 + \gamma_m^2 \Omega^2}}{\Omega^2}\bigg|_{\Omega=\omega_m} = \frac{\gamma_m}{\omega_m}. \tag{2.97}$$

The free-mass SQL can therefore be surpassed by a significant amount of the mechanical quality factor ω_m/γ_m around its resonance frequency. We will explore these two approaches in detail in the next two sections.

2.8 Beating the SQL by Building Correlations

In this section, we will focus on the first approach to beat the SQL, by creating correlations between the shot noise and the back-action noise. This can be achieved by: (i) signal-recycling; (ii) squeezed input; (iii) variational readout. The details of these methods will be discussed.

2.8.1 Signal-Recycling

In the previous section (cf. Sect. 2.6.3), we showed that the signal-recycled interferometer can be mapped into a detuned cavity, in which the amplitude and phase quadratures will rotate and mix with each other. When their fluctuations are reflected

2.8 Beating the SQL by Building Correlations

back to the test mass by the signal-recycling mirror, both will contribute to the radiation-pressure noise, similarly for the shot noise, which will have contributions from both fluctuations. Therefore, the shot noise and radiation-pressure noise naturally acquire correlations. In the case of phase-quadrature readout, it can be shown that this correlation is given by [8]:

$$S_{ZF} = \hbar \frac{\Delta_{\text{eff}}(\Delta_{\text{eff}}^2 + 3\gamma_{\text{eff}}^2 - \Omega^2)}{2\gamma_{\text{eff}}(\gamma_{\text{eff}}^2 + \Omega^2)}. \tag{2.98}$$

This accounts for the two minima in the sensitivity curve, which surpass the SQL, as shown in Fig. 2.15.

2.8.2 Squeezed Input

As pointed out in the pioneering work of [16], a frequency-dependent squeezed input can be used to surpass the SQL; in this case, the shot noise and radiation-pressure noise are naturally correlated as a result of the correlation between the amplitude and phase quadratures of the input squeezed light. To illustrate this point, we will only look at the scheme discussed in Sect. 2.6.2 without a signal-recycling mirror. It can be easily generalized and extended to general schemes. Given a squeezed input, the amplitude and phase quadrature will be transformed according to Eqs. (2.17) and (2.18). As shown in [16], if the squeezing angle ϕ has the following frequency dependence:

$$\phi(\Omega) = -\text{arccot}\,\mathcal{K}(\Omega), \tag{2.99}$$

the amplitude and phase fluctuations are squeezed at low and high frequencies, respectively, and the noise spectrum will be reduced by a overall squeezing factor of e^{2q}, namely:

$$S_{\text{sqz}}^h = e^{-2q}\left[\mathcal{K} + \frac{1}{\mathcal{K}}\right]\frac{h_{\text{SQL}}^2}{2}, \tag{2.100}$$

which surpasses the SQL at around the most sensitive frequencies, as indicated in Fig. 2.18. It is possible to achieve a frequency-dependent squeezing, as prescribed in Eq. (2.99), because different sideband frequencies all represent different degrees of freedom, and therefore squeezing them in different ways is totally allowed. In practice, one must invent the right device that generates such a frequency dependence. This usually implies devices with a certain time scale that is comparable to the detection band of the detector, which is very long compared with usual quantum optical devices (Fig. 2.17).

Kimble et al. invented one such device—a detuned Fabry Pérot cavity. For Fabry-Pérot Michelson interferometers, they showed that a squeezing angle of arctan $\mathcal{K}$ can

Fig. 2.17 A schematic plot of the frequency-dependent squeezed input configuration. A frequency-independent squeezed light, filtered by two detuned Fabry-Pérot cavities in sequence, produces the required frequency dependence

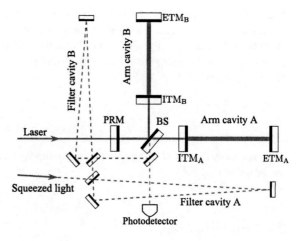

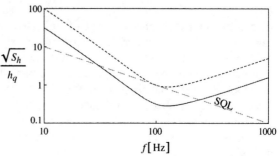

Fig. 2.18 The GW strain-referred sensitivity of a frequency-dependent squeezing input interferometer (*solid curve*). The squeezing factor is $e^{-2q} = 0.1$. The other specification is the same as Fig. 2.12

be generated by feeding a frequency-independent squeezed light into two consecutive Fabry-Pérot cavities, each with a perfect end mirror, and prescribed values of detuning and bandwidth. The specifications for the filter cavity parameters can be obtained by using the technique given in the Appendix of [21]. This technique is valid for general schemes, such as a signal-recycling configuration, shown explicitly in [14], in which the noise spectrum is also reduced by the squeezing factor for the entire detection band.

2.8.3 Variational Readout: Back-Action Evasion

Another natural way to build correlations is to measure a certain combination of the amplitude quadrature $\hat{b}_1$ and phase quadrature $\hat{b}_2$ at the output, namely:

$$\hat{b}_\zeta = \hat{b}_1 \cos \zeta + \hat{b}_2 \sin \zeta$$
$$= (\hat{a}_1 \cos \zeta + \hat{a}_2 \sin \zeta) - \hat{a}_1 \mathcal{K} \sin \zeta + \sqrt{2\mathcal{K}} \frac{h}{h_{\text{SQL}}} \sin \zeta. \quad (2.101)$$

2.8 Beating the SQL by Building Correlations

Fig. 2.19 A schematic plot of the variational-readout configuration. The output is filtered by two detuned Fabry-Pérot cavities before the detection

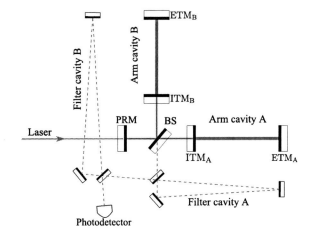

The shot noise ($\hat{a}_1 \cos\zeta + \hat{a}_2 \sin\zeta$) and the radiation-pressure noise $\hat{a}_1 \mathcal{K} \sin\zeta$ has non-zero correlation. If the detection angle ζ has the following frequency dependence:

$$\zeta(\Omega) = \text{arccot}\,\mathcal{K}(\Omega), \quad (2.102)$$

we can completely evade the effect of back-action, and obtain a shot-noise only sensitivity, namely:

$$S_{\text{var}}^h = \frac{h_{\text{SQL}}^2}{2\mathcal{K}}. \quad (2.103)$$

Such back-action- evading scheme was first invented by Vyatchanin et al. in a time domain formalism [24, 25], aimed at detecting GWs with known arrival time. The above frequency-domain formalism was developed by Kimble et al., and it is valid for all possible stationary signals. The required frequency dependency can be achieved in a similar way to that of frequency-dependent squeezing, i.e., by filtering the output through detuned Fabry-Pérot cavity, as shown schematically in Fig. 2.19. The specifications for the filter cavities can also be obtained by using the results in Ref. [21]. The corresponding sensitivity curve is shown in Fig. 2.20.

2.8.4 Optical Losses

In the above discussions, we have assumed an ideal lossless situation. However, in reality, there are multiple channels in which losses can be introduced. These include the scattering and losses in the optical elements, and the non-zero transmission of the end mirrors. These optical losses will introduce uncorrelated vacuum fluctuations,

Fig. 2.20 The GW strain-referred sensitivity of a variational-readout scheme (*solid curve*). The low-frequency back-action noise is completed evaded, thus achieving a sensitivity limited only by shot noise

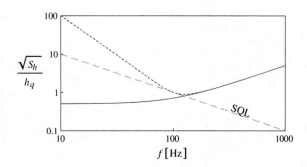

and destroy the quantum coherence. They can be modeled by an overall reduction of the field amplitude by $\sqrt{1-\epsilon}$, and then an introduction of $\sqrt{\epsilon}$ for the vacuum:

$$\vec{b} \to \sqrt{1-\epsilon}\,\vec{b} + \sqrt{\epsilon}\,\vec{n}. \tag{2.104}$$

Here $\vec{n}$ is associated with the uncorrelated vacuum fluctuations. They will not only affect the injected squeezed state but also the output field, which undermines the sensitivity. As shown by Kimble et al., the squeezed-input configuration is reasonably robust against optical loss as long as the filter cavities have a length comparable to the arm cavity (∼4 km). However, the variational-readout scheme is very susceptible to the optical losses. If we apply the condition in Eq. (2.102), the noise spectrum with the output loss [cf. Eq. (2.104)] is:

$$S_{\text{loss}}^h(\Omega) = \left[\frac{\epsilon}{2(1-\epsilon)\mathcal{K}\sin^2\zeta} + \frac{1}{2\mathcal{K}}\right]h_{\text{SQL}}^2. \tag{2.105}$$

At low frequencies, $\mathcal{K} \sim \Omega^{-2}$ and $\sin^2\zeta \sim \Omega^4$, and thus the first term scales as Ω^{-2}. This means that the loss will severely affect the low-frequency sensitivity. In addition, as shown by Kimble et al., around the frequency where the SQL is attained for a conventional scheme without a signal-recycling mirror, the SQL beating ratio of a variational-readout scheme with loss is given by:

$$\mu \equiv \sqrt{\frac{S_{\text{SQL}}^h}{S_{\text{var}}^h}} \approx \sqrt[4]{\epsilon}, \tag{2.106}$$

only if the interferometer can manage a factor of $1/\sqrt{\epsilon}$ times stronger optical power. Given a typical loss of 0.01, this produces a factor of 0.3 and a power 10 times greater, which is rather challenging. Therefore, a low-loss optical setup is essential for implementation.

2.9 Optical Spring: Modification of Test-Mass Dynamics

Apart from building correlations, as shown by Eq. (2.97), the free-mass SQL can also be surpassed by modifying the dynamics of the test-mass. One might expect that this will require a significant modification of the topology of current GW detectors. As it turns out, a detuned signal-recycling interferometer naturally achieves this. This is intimately connected to the rotation of the amplitude and phase quadrature in the signal-recycling cavity, which produces correlations between the shot noise and radiation-pressure noise. As we have previously mentioned, the radiation pressure force not only depends on the amplitude quadrature, but also on the phase quadrature. Since the latter contains the test-mass displacement, it induces a position-dependent force and modifies the test-mass dynamics; This phenomenon is also called the "optical-spring" effect. A similar idea, but with a different configuration, was first proposed by [19], and is the so-called "optical bar" GW detector.

2.9.1 Qualitative Understanding of Optical-Spring Effect

The optical-spring effect can be understood qualitatively by looking at the case of a single detuned cavity. The displacement of the end mirror (test-mass) x will change the intra-cavity power I_c, which in turn changes the radiation-pressure force. In the adiabatic limit, the intra-cavity power as a function of x reads:

$$I_c(x) = \frac{\gamma^2 I_c^{\max}}{\gamma^2 + [\Delta + (\omega_0 x/L)]^2}, \qquad (2.107)$$

which is shown in the left panel of Fig. 2.21. Since the radiation-pressure force is equal to $F(x) = I_c(x)/c$, such a position-dependent force will introduce a rigidity, which is minus the derivative of the force $-dF(x)/dx$, around the equilibrium point $x = 0$. Depending on the sign of the detuning, it will create either negative or positive rigidity. In the case of a detuned signal-recycling configuration for AdvLIGO, a strong optical-spring effect can shift the pendulum frequency of 1 Hz up to the detection band. Around the new resonant frequency, we can surpass the free-mass SQL. This actually accounts for the low-frequency dip in Fig. 2.15. One can refer to Ref. [7] for a detailed discussion of the mechanical resonance due to this optical-spring effect.

Due to a delayed response of the intra-cavity power to the test-mass motion, the optical spring also has a friction component. For $\Delta < 0$, such a delayed response gives a positive damping, which pumps energy out of the test mass. For $\Delta > 0$, the damping is negative, which will destabilize the system. Using the input–output relation derived in Sect. 2.5.3, one can easily determine the expression for an optical spring in the frequency domain as:

$$K(\Omega) = -\frac{2I_c\omega_0}{Lc}\frac{\Delta}{(\Omega - \Delta + i\gamma)(\Omega + \Delta + i\gamma)}. \qquad (2.108)$$

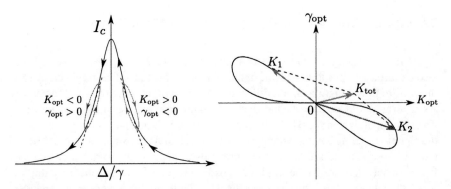

Fig. 2.21 The optical power as a function of cavity detuning Δ (*left*). A double optical spring, $K_{\text{tot}}(\Omega) = K_1(\Omega) + K_2(\Omega)$, which has both positive rigidity and damping (*right*). The *black* curve (*right*) shows a parametric plot of the optical spring $K(\Omega)$ as a function of detuning

For sideband frequencies $\Omega < \Delta, \gamma$, we can perform a Taylor expansion and obtain

$$K(\Omega) = \frac{2I_c \omega_0}{Lc} \left[\frac{\Delta}{\gamma^2 + \Delta^2} + \frac{2i\gamma\Delta}{(\gamma^2 + \Delta^2)^2} \Omega \right] \equiv K_{\text{opt}} - i\gamma_{\text{opt}}\Omega. \quad (2.109)$$

We have introduced the rigidity K_{opt} (real part of K) and the damping γ_{opt} (imaginary part). As we can see, the positive (negative) rigidity is always accompanied by a negative (positive) damping. In either case, the system is unstable. To stabilize the system, one can use a feedback control as described by [6]. An interesting alternative is to implement the idea of a double optical spring, by pumping the cavity with two lasers at different frequencies [12, 22]. One laser with a small detuning provides a large positive damping while another with a large detune, but with a high power, provides a strong restoring force. The resulting system is self-stabilized with both positive rigidity and positive damping, as shown schematically in the right panel of Fig. 2.21.

2.10 Continuous State Demolition: Another Viewpoint on the SQL

In the previous section (cf. Sect. 2.7), we derive the SQL by focusing on the quantum nature of the detection (the optical field). In this section, we will derive the SQL from another perspective—continuous state demolition. This will guide us to finding new approaches to surpassing the SQL.

In the Heisenberg picture, suppose we attempt to measure the position of a free mass successively at discrete times separated by τ. We measure $\hat{x}$ at time t_1, *then right after* t_1, the quantum state of the test mass is characterized by a standard deviation comparable to the individual-measurement error ϵ, or:

2.10 Continuous State Demolition: Another Viewpoint on the SQL

$$\Delta x(t_1 + 0) = \epsilon. \tag{2.110}$$

The value of ϵ decreases indefinitely as individual-measurement sensitivity increases. Applying free-mass quantum mechanics for the duration of $t \in (t_1, t_2)$, we have

$$[\hat{x}(t_1 + 0), \hat{x}(t_2 - 0)] = \frac{i\hbar(t_2 - t_1)}{m} = \frac{i\hbar\tau}{m}. \tag{2.111}$$

The decreasing ϵ will lead to an increasing variance in $\Delta x(t_2 - 0)$ right before the second measurement. This is because the Heisenberg Uncertainty Principle dictates that:

$$\Delta x(t_1 + 0) \cdot \Delta x(t_2 - 0) \geq \hbar\tau/(2m), \tag{2.112}$$

and

$$\Delta x(t_2 - 0) > \frac{\hbar\tau}{2m\epsilon}. \tag{2.113}$$

This large variance in the $x(t_2 - 0)$ distribution will be reduced down to ϵ by the subsequent measurement on $x(t_2)$—but at the price of *demolishing* a quantum state with a large spread of $x(t_2)$ into classical superpositions of quantum states with much smaller variances.

If the successive measurements are done without coordination, i.e., if the *meters* that collapse the mirror's states at t_1 and t_2 are not correlated, then the demolition will cause an additional noise, because the new center of the wavefunction after collapse is randomly chosen among a distribution with variance $\Delta x(t_2 - 0)$. If we now characterize the noise of each individual measurement in position, we obtain,

$$\Delta x \geq \max\left(\epsilon, \frac{\hbar\tau}{2m\epsilon}\right) \geq \sqrt{\frac{\hbar\tau}{2m}}. \tag{2.114}$$

This provides us with the scale of the standard quantum limit. In fact, if for any pulse with duration τ, which can vary at all scales, our measurement error is always:

$$\Delta x \sim \sqrt{\frac{\hbar\tau}{2m}}, \tag{2.115}$$

then the *noise spectral density* of the device is characterized by:

$$S^x(\Omega) \sim \frac{\hbar}{2m\Omega^2}. \tag{2.116}$$

Therefore, the SQL can be traced back to the fact that the test mass positions at different times do not commute with themselves [cf. Eq. (2.111)].

2.11 Speed Meters

This alternative viewpoint on the SQL naturally brings us to the idea of a speed meter, which measures the speed (momentum) instead of the position of the test mass. Since the momentum of a free mass is a conserved dynamical quantity, and its Heisenberg operators at different times commute with each other, one can measure it continuously without imposing additional noise, thus allowing us to surpass the SQL.

Apart from beating the SQL, other more practical issues that make the speed meter attractive are the following: (1) the previous schemes require frequency-dependent squeezing or readout which has led to the requirement of two extra kilometer-scale filter cavities—a high cost in practical construction; (2) the high value of $\mathcal{K}$ at low frequencies leads to a strong sensitivity to optical losses. As it turns out, these issues are resolved simultaneously when we try to build a speed meter, which has a constant sensitivity to the speed of test masses in low frequencies—the requirement for frequency dependence on squeezing and readout quadrature therefore vanishes. In fact, if we imagine an interferometer with a constant κ [cf. Eq. (2.29)] or $\mathcal{K}$ [cf. Eq. (2.41)] in the input–output relation, it will indeed respond to the speed of the mirror.

Here we will discuss two realizations, both found as prototypes in the early papers of [5, 15], but later gradually deformed into the shape of kilometer-scale laser interferometers [11, 20, 21].

2.11.1 Realization I: Coupled Cavities

A possible Michelson variant is shown in Fig. 2.22. An additional sloshing cavity is added after the interferometer output. It has an input-mirror with transmissivity T_s and a totally-reflected end mirror. There is another extraction mirror (with a transmissivity of T_0) between the interferometer output and the sloshing cavity, through which we read out the signal. This configuration emerges from the two-resonator model of Braginsky and Khalili, where it was pointed out that if two resonators are coupled, then a sloshing of signal light between the two cavities cancel each other out, leaving only a sensitivity to the change in mirror position, i.e., the speed. The characteristic sloshing frequency is given by:

$$\omega_s = \frac{\sqrt{T_s} c}{2L}. \tag{2.117}$$

The explicit expression for the response function of the output to the test mass motion can be found by analyzing the input–output relation, giving:

$$T(\Omega) = \frac{\omega_0}{L} \frac{i\Omega}{\Omega^2 - \omega_s^2 + i\gamma\Omega}, \tag{2.118}$$

2.11 Speed Meters

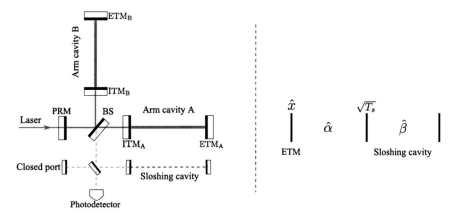

Fig. 2.22 A schematic plot of a speed-meter configuration with coupled cavities. A sloshing cavity is added after the *dark* port. Such a configuration can be mapped into a three-mirror cavity as schematically shown in the *right*-hand panel, forming a pair of coupled cavities

with $\gamma \equiv 4T_0 c/L$ due to the extraction mirror. This can also be derived by fitting the speed meter into the two-resonator model. We can map the speed meter configuration into a three-mirror cavity, as shown schematically in Fig. 2.22. The '$\hat{x}$' corresponds to the differential motion of the end test masses. The left cavity corresponds to the power-recycled Michelson interferometer, and the right cavity is still the sloshing cavity. The optical fields are summarized by two cavity modes $\hat{\alpha}$ and $\hat{\beta}$. Their equations of motion are:

$$\dot{\hat{\alpha}}(t) + \gamma\, \hat{\alpha}(t) = i\omega_s\, \hat{\beta}(t) + i\frac{\omega_0}{L}\hat{x}(t), \qquad (2.119)$$

$$\dot{\hat{\beta}}(t) = i\omega_s\, \hat{\alpha}(t). \qquad (2.120)$$

We have assumed that both cavity modes are on resonance with respect to the carrier laser frequency. In addition, we attribute the decay only to the left cavity (due to the extraction mirror). The coupling between the two modes is manifested by the sloshing terms on the right hand side (proportional to ω_s). Solving the above equation in the frequency domain will immediately give the result in Eq. (2.118).

The GW strain sensitivity of such a configuration can be derived by using the input–output formalism that we have introduced. Given a frequency-independent squeezing (phase squeezing factor e^{-2q}), the result is (refer to Ref. [21] for more details):

$$S^h = \left[\frac{e^{-2q}}{2\mathcal{K}_{\rm sm}} + \frac{e^{2q}(\cot\zeta - \mathcal{K}_{\rm sm})^2}{2\mathcal{K}_{\rm sm}}\right] h_{\rm SQL}^2, \qquad (2.121)$$

where ζ is the readout quadrature angle, and

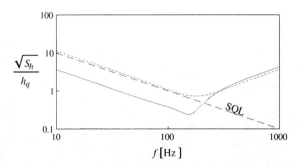

Fig. 2.23 The GW strain-referred sensitivity of the speed meter scheme shown in Fig. 2.22. We have chosen $\gamma/2\pi = 210\,\text{Hz}$, $\omega_s/2\pi = 180\,\text{Hz}$ and $I_c = 800\,\text{kW}$. The *dotted* curve is the case without a phase-squeezed input, and the *solid* curve has $e^{-2q} = 0.1$ (a 10 dB squeezing)

$$\mathcal{K}_{\text{sm}} = \frac{16\omega_0 \gamma I_c}{mcL[(\Omega^2 - \omega_s^2)^2 + \gamma^2 \Omega^2]}. \tag{2.122}$$

The first term in S^h is the shot-noise term, and the second is the radiation-pressure noise. At low frequencies, $\mathcal{K}_{\text{sm}}$ is almost a constant. By choosing $\cot \zeta = \mathcal{K}_{\text{sm}}(0)$, the low-frequency radiation-pressure (back-action) can be completely evaded. We show the resulting GW strain sensitivity in Fig. 2.23.

There is, however, a subtle issue: the original Braginsky and Khalili argument stated that momentum can be measured without additional noise—yet in speed meters, we still need back-action evasion: is this still consistent? The answer is yes, because when speed is coupled to an external observable, it ceases to be proportional to the conserved, canonical momentum. Here, we sketch a mathematical proof by Khalili. Suppose the Lagrangian is:

$$\mathcal{L} = \frac{1}{2}m\dot{x}^2 + \alpha \dot{a}_1 x, \tag{2.123}$$

which is the model of a speed meter: the time derivative of the external observable a_1 represents the sloshing. Here, the quantity α is a coupling constant. This is equivalent to:

$$\mathcal{L} = \frac{1}{2}m\dot{x}^2 - \alpha a_1 \dot{x}, \tag{2.124}$$

where the coupling becomes a speed coupling, but the canonical momentum is given by:

$$p = \frac{\partial \mathcal{L}}{\partial \dot{x}} = m\dot{x} - \alpha a_1, \tag{2.125}$$

which is conserved, but differs from the kinetic momentum, $m\dot{x}$. This means that a meter that measures speed must add a constant $\alpha a_1/m$, in order to evade the back-action. Note that this is a constant combination between the output phase quadrature

2.11 Speed Meters

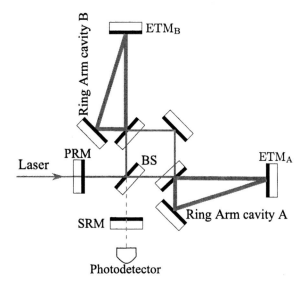

Fig. 2.24 A schematic plot of a Sagnac-type speed meter configuration

and amplitude quadrature—therefore, back-action evasion is straightforward without requiring additional filtering. This is manifested by the fact that $\mathcal{K}_{sm}$ is almost constant at low frequencies.

2.11.2 Realization II: Zero-Area Sagnac

Another possible speed meter configuration is a zero-area Sagnac interferometer, which is shown schematically in Fig. 2.24. This has a different optical topology from the Michelson scheme, where the light travels through two opposite loops in the interferometer. To understand its response to the test-mass motion, we can look at a single trip. The light propagating towards arm A first picks up a phase shift proportional to ETM_A, displacement $\hat{x}_A(t)$, and it accumulates another phase shift due to motion of ETM_B, but at $t - \tau$ (τ is the delay time). A similar situation holds for the light propagating towards arm B, but with the roles of ETM_A and ETM_B swapped. When they recombine at the beam splitter, the total phase shift is simply:

$$\phi_{\text{tot}}(t) \propto \hat{x}_A(t) + \hat{x}_B(t - \tau) - \hat{x}_B(t) - \hat{x}_A(t - \tau) \approx [\dot{\hat{x}}_A(t) - \dot{\hat{x}}_B(t)]\tau. \quad (2.126)$$

It naturally has no response to a static change in arm length, but only to the differential speed of the two test masses. Therefore, it is a natural speed meter. This has been recognized by the GW community, but connection with the "Quantum Nondemolition" (QND) speed meter, has never been made. As shown in [25], the GW strain sensitivity of such a configuration is identical to the previous coupled-cavities configuration.

2.12 Conclusions

We have introduced the basic concepts for studying the quantum dynamics and associated quantum noise of an interferometric GW detector. Different insights into the origin of the SQL allow us to find out possible approaches towards surpassing it. In particular, we have considered modifying the input/ouput optics using either frequency-dependent squeezing or a variational readout, and modifying the test mass dynamics through the optical spring effect; and measuring a QND observable of the test mass by using a speed meter. This not only serves as a review of advanced configurations for future GW detectors, but it additionally provides valid examples to help clarify many subtle issues in continuous quantum measurement.

References

1. K.J. Blow, R. Loudon, S.J.D. Phoenix, T.J. Shepherd, Continuum fields in quantum optics. Phys. Rev. A **42**, 4102 (1990)
2. V.B. Braginsky, Classical and Quantum Restrictions on the Detection of Weak Disturbances of a Macroscopic Oscillator. JETP **26**, 831 (1968)
3. V.B. Braginsky, F.Y. Khalili, *Quantum Measurement* (Cambridge University Press, Cambridge, 1992)
4. V.B. Braginsky, M.L. Gorodetsky, F.Y. Khalili, Optical bars in gravitational wave antennas. Phys. Lett. A **232**, 340 (1997)
5. V.B. Braginsky, M.L. Gorodetsky, F.Y. Khalili, K.S. Thorne, Noise in gravitational-wave detectors and other classical-force measurements is not influenced by test-mass quantization. Phys. Rev. D **61**, 044002 (2000)
6. A. Buonanno, Y. Chen, Quantum noise in second generation, signalrecycled laser interferometric gravitational-wave detectors. Phys. Rev. D **64**, 042006 (2001)
7. A. Buonanno, Y. Chen, Signal recycled laser-interferometer gravitational-wave detectors as optical springs. Phys. Rev. D **65**, 042001 (2002)
8. A. Buonanno, Y. Chen, Scaling law in signal recycled laserinterferometer gravitational-wave detectors. Phys. Rev. D **67**, 062002 (2003)
9. C.M. Caves, B.L. Schumaker, New formalism for two-photon quantum optics. I - Quadrature phases and squeezed states. II - Mathematical foundation and compact notation. Phys. Rev. A **31**, 3068 (1985)
10. C.M. Caves, K.S. Thorne, R.W. Drever, V.D. Sandberg, M. Zimmermann, On the measurement of a weak classical force coupled to a quantummechanical oscillator. I. Issues of principle. Rev. Mod. Phys. **52**, 341 (1980)
11. Y. Chen, Sagnac interferometer as a speed-meter-type, quantumnondemolition gravitational-wave detector. Phys. Rev. D **67**, 122004 (2003)
12. T. Corbitt, Y. Chen, E. Innerhofer, H. Mueller-Ebhardt, D. Ottaway, H. Rehbein, D. Sigg, S. Whitcomb, C. Wipf, N. Mavalvala, An All-Optical Trap for a Gram-Scale Mirror. Phys. Rev. Lett. **98**, 150802–150804 (2007)
13. R.W.P. Drever, *Gravitational Radiation* (North-Holland, Amsterdam, 1983)
14. J. Harms, Y. Chen, S. Chelkowski, A. Franzen, H. Vahlbruch, K. Danzmann, R. Schnabel, Squeezed-input, optical-spring, signal-recycled gravitational-wave detectors. Phys. Rev. D **68**, 042001 (2003)

References

15. F.Y. Khalili, Y. Levin, Speed meter as a quantum nondemolition measuring device for force. Phys. Rev. D **54**, 004735 (1996)
16. H.J. Kimble, Y. Levin, A.B. Matsko, K.S. Thorne, S.P. Vyatchanin, Conversion of conventional gravitational-wave interferometers into quantum nondemolition interferometers by modifying their input and/or output optics. Phys. Rev. D **65**, 022002 (2001)
17. B.J. Meers, Recycling in laser-interferometric gravitational-wave detectors. Phys. Rev. D **38**, 2317 (1988)
18. J. Mizuno, Comparison of Optical Configurations for Laser-Interferometric Gravitational-Wave Detectors. PhD thesis, Max-Planck Institut für Quantenoptik, Garching, Germany, 1995
19. J. Mizuno, K.A. Strain, P.G. Nelson, J.M. Chen, R. Schilling, A. Rüdiger, W. Winkler, K. Danzmann, Resonant sideband extraction: a new configuration for interferometric gravitational wave detectors. Phys. Lett. A **175**, 273 (1993)
20. P. Purdue, Analysis of a quantum nondemolition speed-meter interferometer. Phys. Rev. D **66**, 022001 (2002)
21. P. Purdue, Y. Chen, Practical speed meter designs for quantum nondemolition gravitational-wave interferometers. Phys. Rev. D **66**, 122004 (2002)
22. H. Rehbein, H. Müller-Ebhardt, K. Somiya, S.L. Danilishin, R. Schnabel, K. Danzmann, Y. Chen, Double optical spring enhancement for gravitational-wave detectors. Phys. Rev. D **78**, 062003 (2008)
23. J.Y. Vinet, B. Meers, C. Man, A. Brillet, Optimization of long-baseline optical interferometers for gravitational-wave detection. Phys. Rev. D **38**, 433 (1988)
24. S.P. Vyatchanin, A.B. Matsko, Quantum limit on force measurements. JETP **77**, 218 (1993)
25. S.P. Vyatchanin, E.A. Zubova, Quantum variation measurement of a force. Phys. Lett. A **201**, 269–274 (1995)

Chapter 3
Modifying Input Optics: Double Squeezed-Input

3.1 Preface

In this chapter, we consider improving the sensitivity of future interferometric GW detectors by modifying their input optics. In particular, we discuss simultaneously injecting two squeezed light, filtered through a resonant Fabry-Pérot cavity, into the dark port of the interferometer. This is motivated by the work of Corbitt et al. [29], in which a similar scheme, but with a single squeezed light was proposed and analyzed. Here we show that the extra squeezed light, together with an additional homodyne detection suggested previously by Khalili [9], allows a reduction of quantum noise over the entire detection band. To motivate future implementations, we take into account a realistic technical noise budget for the Advanced LIGO (AdvLIGO), and numerically optimize the parameters of both the filter and the interferometer for detecting gravitational-wave signals from two important astrophysics sources: neutron-star–neutron-Star (NS–NS) binaries and Bursts. Assuming the optical loss of the ~ 30 m filter cavity to be 10 ppm per bounce, and 10dB squeezing injection, the corresponding quantum noise with optimal parameters decreases by a factor of 10 at high frequencies and goes below the technical noise at low and intermediate frequencies. This is a joint research by Farid Khalili, Yanbei Chen and myself. It is published in Phys. Rev. D **80**, 042006 (2009).

3.2 Introduction

During the last decade, several laser interferometric gravitational-wave (GW) detectors including LIGO [10], VIRGO [17], GEO600 [16] and TAMA [19] have been built and operated almost at their design sensitivity, aiming at extracting GW signals from various astrophysical sources. At present, the development of next-generation detectors, such as AdvLIGO [20], is also under way, and the sensitivities of these advanced detectors are anticipated to be limited by quantum noise over almost the whole observational band from 10 Hz to 10^4 Hz. At high frequencies, the dominant quantum noise is *photon shot noise*, caused by phase fluctuation of the optical field; while at low frequencies, the *radiation-pressure noise*, due to the amplitude fluctuation, dominates and it exerts a noisy random force on the probe masses. These two noises, if uncorrelated, will impose a lower bound on the noise spectrum, which is called the Standard Quantum Limit (SQL). In terms of the GW strain $h \equiv \Delta L/L$, this limit is given by

$$S_h^{\text{SQL}} = \frac{8\hbar}{m\Omega^2 L^2}. \tag{3.1}$$

This can also be derived from the fact that position measurements of the free test mass do not commute with themselves at different times [21].

The existence of the SQL was first realized by Braginsky in the 1960s [9, 14]. Since then, various approaches have been proposed to beat the SQL. One recognized by Braginsky is to measure conserved quantities of the probe masses, also called Quantum Nondemolition (QND) quantities. This can be achieved, e.g. by adopting "speed meter" configurations [1, 5, 15, 25, 35, 36], which measure the conserved quantity—momentum rather than the position. An alternative is to change the dynamics of the probe mass, e.g. by using optical rigidity [3, 6], in which case the above-mentioned free mass SQL is no longer relevant. As shown by Buonanno and Chen [4, 7, 26], optical rigidity exists in signal-recycled (SR) interferometric GW detectors; therefore we can beat the SQL without radical redesigns of the existing topology of the interferometers. Another approach is to modify the input and/or output optics of the interferometers, such that *photon shot noise* and *radiation-pressure noise* are correlated. After the initial paper by Unruh [18], this was further developed by other authors [2, 11, 23, 27, 29, 30, 31, 34, 37]. A natural way to achieve this is injecting a squeezed vacuum state, whose phase and amplitude fluctuations are correlated, into the dark port of the interferometer. With great advancements in preparation of the squeezed state [33, 12], squeezed-input interferometers will be promising candidates for third-generation GW detectors. As elaborated in the work of Kimble et al. [2], frequency-dependent squeezing is essential to reduce the quantum noise at various frequencies of the observation band. In addition, they demonstrated that this can be realized by filtering the frequency-independent squeezed vacuum state through two detuned Fabry-Pérot cavities before sending into the interferometer. Their results were extended by Purdue and Chen [15] who discussed the filters for general cases.

Another method, which also uses an additional filter cavity and squeezed state, but in a completely different way, was proposed by Corbitt, Mavalvala, and Whitcomb (hereafter referred to as CMW) [29]. They proposed to use a tuned optical cavity as a high-pass filter for the squeezed state. This scheme does not create the noises correlation, but instead, renders their spectral densities frequency-dependent. At high frequencies, the phase-squeezed vacuum state gets reflected by the filter and enters the interferometer such that high-frequency shot noise is reduced; while at low frequencies, ordinary vacuum transmits through the filter and enters the interferometer, thus low-frequency radiation-pressure noise remains unchanged. One significant

3.2 Introduction

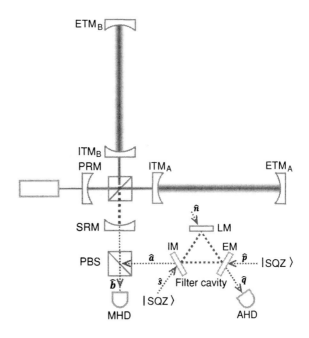

Fig. 3.1 Schematic plot of the proposed configuration. Two squeezed light beams $\hat{s}, \hat{p}$ are injected from both side of the filter cavity rather than the one which was considered in Refs. [29, 9]. The signal is detected by the main homodyne detector (*MHD*) and an additional detector (*AHD*) is placed at the idle port of the filter cavity. From Ref. [24]

advantage is that the squeezed vacuum state does not really enter the filter cavity, and thus it is less susceptible to the optical losses. However, it does not perform so well as hoped, and there is a noticeable degradation of sensitivity in the intermediate-frequency range. One of us—Khalili [9] pointed out that this has to do with the quantum entanglement between the optical fields at the two ports of the filter cavity. Equivalently, it can be interpreted physically as follows: some information about the phase and amplitude fluctuations flows out from the idle port of the filter cavity and the remaining quantum state which enters the interferometer is not pure. In order to recover the sensitivity, the filter cavity needs to have a low optical loss, such that this information can be collected by an additional homodyne detector (AHD) at the idle port. Given an achievable optical loss of the filter cavity $\sim$10 ppm per bounce, Khalili showed that we can obtain the desired sensitivity at intermediate frequencies. A natural extension of this scheme is to send an additional squeezed vacuum state into the idle port of the filter cavity, such that the low-frequencies radiation-pressure noise is also suppressed. The corresponding configuration is shown schematically in Fig. 3.1, where two squeezed vacuum states $\hat{s}$ and $\hat{p}$ are injected from two ports of the filter cavity, and some ordinary vacuum state $\hat{n}$ leaks into the filter due to optical losses. By optimizing the squeezing angles of these squeezed states, we will show that the resulting quantum noise is reduced over the entire observational band.

The outline of this chapter is as follows: in Sect. 3.3, we will calculate the quantum noises in this double squeezed-input CMW scheme with AHD (later referred to as CMWA). We will use the same notation as in Ref [26], which enables us to extend the results in Ref. [9] to the case of signal-recycled interferometers easily.

Table 3.1 Main quantities in this paper

Quantity	Value for estimates	Descriptions
Ω		GW frequency
c	3.0×10^8 m/s	Speed of light
ω_0	1.8×10^{15} s^{-1}	Laser frequency
m	40 kg	Test mass
L	4 km	Arm cavity length
I_c	840 kW	Optical power
$\iota_c = \frac{8\omega_0 I_c}{mLc}$	$(2\pi \times 100)^3$ s^{-3}	
γ_{arm}	$2\pi \times 100$ s^{-1}	Cavity bandwith
r_{SR}		SRM reflectivity
ϕ_{SR}		Phase detuning of SR cavity
δ		Effective detuning
γ		Effective bandwidth
ϕ		MHD homodyne angle
ζ		AHD homodyne angle
L_f	30 m	Filter cavity Length
$n_{\text{I,E,L}} = \frac{T_{\text{I,E,L}}^2}{2\tau_f}$		
$\gamma_f = \gamma_{\text{I}} + \gamma_{\text{E}} + \gamma_{\text{L}}$		Filter cavity bandwith
$r_i (i = s, p)$	$(\ln 10)/2$ (10dB)	Squeezing factors
$\theta_i (i = s, p)$		Squeezing angles

In Sect. 3.4, we numerically optimize the parameters of this new scheme, as employed in the search for GW signals from NS–NS binaries and Bursts. Finally, we summarize our results in Sect. 3.5 For simplicity, we will neglect the optical losses inside the main interferometer, but we do consider the losses from the filter cavity, and also from the non-unity quantum efficiency of the photodiodes. The losses from the main interferometer are not expected to be important, as shown in Refs. [2, 7, 9]. The main quantities used in this paper are listed in Table 3.1.

3.3 Quantum Noise Calculation

3.3.1 Filter Cavity

In this section, we will derive single-sided spectral densities of the two outgoing fields $\hat{a}$ and $\hat{q}$ as shown in Fig. 3.1. From the continuity of optical fields, we can relate them to the ingoing fields, which include two squeezed vacuum states $\hat{s}$, $\hat{p}$ and one ordinary vacuum $\hat{n}$ entering from the lossy mirror (*LM*) [9]. Specifically, we have

$$\hat{a}(\Omega) = \mathcal{R}_{\text{I}}(\Omega)\hat{s}(\Omega) + \mathcal{T}(\Omega)\hat{p}(\Omega) + \mathcal{A}_{\text{I}}(\Omega)\hat{n}(\Omega), \quad (3.2)$$

$$\hat{q}(\Omega) = \mathcal{R}_{\text{E}}(\Omega)\hat{p}(\Omega) + \mathcal{T}(\Omega)\hat{s}(\Omega) + \mathcal{A}_{\text{E}}(\Omega)\hat{n}(\Omega). \quad (3.3)$$

3.3 Quantum Noise Calculation

Here $\hat{a} = (\hat{a}_A, \hat{a}_\varphi)^T, \hat{q} = (\hat{q}_A, \hat{q}_\varphi)^T$ are amplitude and phase quadratures (subscript A stands for amplitude and φ for phase). In our case, the carrier light beam is resonant inside the filter cavity. Therefore, the effective amplitude reflectivity $\mathcal{R}$, transmissivity $\mathcal{T}$ and loss $\mathcal{A}$ can be written as,

$$\mathcal{R}_I(\Omega) = \frac{\gamma_I - \gamma_E - \gamma_L + i\Omega}{\gamma_f - i\Omega}, \quad \mathcal{R}_E(\Omega) = \frac{\gamma_E - \gamma_I - \gamma_L + i\Omega}{\gamma_f - i\Omega}, \quad (3.4a)$$

$$\mathcal{T}(\Omega) = \frac{-2\sqrt{\gamma_I \gamma_E}}{\gamma_f - i\Omega}, \quad (3.4b)$$

$$\mathcal{A}_I(\Omega) = \frac{-2\sqrt{\gamma_I \gamma_L}}{\gamma_f - i\Omega}, \quad \mathcal{A}_E(\Omega) = \frac{2\sqrt{\gamma_E \gamma_L}}{\gamma_f - i\Omega}, \quad (3.4c)$$

where $\gamma_f \equiv \gamma_I + \gamma_E + \gamma_L$. They satisfy the following identities:

$$|\mathcal{R}_I(\Omega)|^2 + |\mathcal{T}(\Omega)|^2 + |\mathcal{A}_I(\Omega)|^2 = |\mathcal{R}_E(\Omega)|^2 + |\mathcal{T}(\Omega)|^2 + |\mathcal{A}_E(\Omega)|^2 = 1, \quad (3.5a)$$

$$\mathcal{R}_I^*(\Omega)\mathcal{T}(\Omega) + \mathcal{R}_E(\Omega)\mathcal{T}^*(\Omega) + \mathcal{A}_I^*(\Omega)\mathcal{A}_E(\Omega) = 0. \quad (3.5b)$$

If the input and end mirrors of the filter cavity are identical, namely $\gamma_I = \gamma_E$, we will have $\mathcal{R}_I, \mathcal{R}_E \sim 1$ and $\mathcal{T} \sim 0$ when $\Omega \gg \gamma_f$ and $\mathcal{R}_I, \mathcal{R}_E \sim 0$ and $\mathcal{T} \sim 1$ when $\Omega \ll \gamma_f$. Therefore, the squeezed vacuum $\hat{s}$ enters the interferometer at high frequencies while $\hat{p}$ becomes significant mostly at low frequencies. By adjusting the squeezing factor and angle of these two squeezed fields, we can reduce both the high-frequency shot noise, and the low-frequency radiation-pressure noise, simultaneously.

To calculate the noise spectral densities, we assume that these two squeezed vacuum sates have frequency-independent squeezing angles $\theta_i (i = s, p)$, and can be represented as follows:

$$\hat{s} = \tilde{R}(r_s, \theta_s)\hat{v}_s, \quad \hat{p} = \tilde{R}(r_p, \theta_p)\hat{v}_p, \quad (3.6)$$

with

$$\tilde{R}(r, \theta) \equiv \begin{pmatrix} \cosh r - \cos\theta \sinh r & -\sin\theta \sinh r \\ -\sin\theta \sinh r & \cosh r + \cos\theta \sinh r \end{pmatrix}. \quad (3.7)$$

Here r is the squeezing factor, and $\hat{v}_i = (\hat{v}_{iA}, \hat{v}_{i\varphi})^T$ are ordinary vacuums with single-sided spectral densities: $S_A(\Omega) = S_\varphi(\Omega) = 1$, $S_{A\varphi}(\Omega) = 0$ [19].

The noise spectral densities of the two filter-cavity outputs can then be written as

$$\tilde{S}_{\hat{a}\hat{a}}(\Omega) = |\mathcal{R}_I(\Omega)|^2 \tilde{R}(2r_s, 2\theta_s) + |\mathcal{T}(\Omega)|^2 \tilde{R}(2r_p, 2\theta_p) + |\mathcal{A}_I(\Omega)|^2 \tilde{I}, \quad (3.8)$$

$$\tilde{S}_{\hat{q}\hat{q}}(\Omega) = |\mathcal{R}_E(\Omega)|^2 \tilde{R}(2r_p, 2\theta_p) + |\mathcal{T}(\Omega)|^2 \tilde{R}(2r_s, 2\theta_s) + |\mathcal{A}_E(\Omega)|^2 \tilde{I}, \quad (3.9)$$

$$\widetilde{S}_{\hat{a}\hat{q}}(\Omega) = \mathcal{R}_{\mathrm{E}}(\Omega)\mathcal{T}^*(\Omega)\widetilde{R}(2r_p, 2\theta_p) + \mathcal{R}_{\mathrm{I}}^*(\Omega)\mathcal{T}(\Omega)\widetilde{R}(2r_s, 2\theta_s) + \mathcal{A}_{\mathrm{E}}(\Omega)\mathcal{A}_{\mathrm{I}}^*(\Omega)\widetilde{\mathrm{I}},$$
(3.10)

where $\widetilde{S}_{\hat{a}\hat{q}}(\Omega)$ is the cross correlation between two outputs; $\widetilde{\mathrm{I}}$ is the identity matrix and

$$\widetilde{S}_i(\Omega) = \begin{pmatrix} S_{A,i}(\Omega) & S_{A\varphi,i}(\Omega) \\ S_{A\varphi,i}(\Omega) & S_{\varphi,i}(\Omega) \end{pmatrix},$$
(3.11)

with $i = \hat{a}\hat{a}, \hat{q}\hat{q}, \hat{a}\hat{q}$, whose elements are single-sided spectral densities.

3.3.2 Quantum Noise of the Interferometer

According to Ref. [26], the input–output relation, which connects the ingoing fields $\hat{a}$ and the GW signal h with the outgoing fields $\hat{b}$, for a signal-recycled interferometer can be written as:

$$\hat{b} = \frac{1}{M}\left(\widetilde{C}\hat{a} + \mathbf{D}\frac{h}{h_{\mathrm{SQL}}}\right).$$
(3.12)

In the above equation,

$$M = [\delta^2 - (\Omega + i\gamma)^2]\Omega^2 - \delta\iota_c,$$
(3.13)

and $\widetilde{C}$ is the transfer function matrix, with elements:

$$\widetilde{C}_{11} = \widetilde{C}_{22} = \Omega^2(\Omega^2 - \delta^2 + \gamma^2) + \delta\iota_c,$$
(3.14a)

$$\widetilde{C}_{12} = -2\gamma\delta\Omega^2, \quad \widetilde{C}_{21} = 2\gamma\delta\Omega^2 - 2\gamma\iota_c,$$
(3.14b)

where $\iota_c \equiv 8\omega_0 I_c/(mcL)$. The elements of the transfer function vector $\mathbf{D}$ are

$$D_1 = -2\delta\sqrt{\gamma\iota_c}\Omega, \quad D_2 = -2(\gamma - i\Omega)\sqrt{\gamma\iota_c}\Omega.$$
(3.15a)

The effective detuning δ, and bandwidth γ, are given by:

$$\delta = \frac{2r_{\mathrm{SR}}\gamma_{\mathrm{arm}}\sin(2\phi_{\mathrm{SR}})}{1 + r_{\mathrm{SR}}^2 + 2r_{\mathrm{SR}}\cos(2\phi_{\mathrm{SR}})},$$
(3.16a)

$$\gamma = \frac{(1 - r_{\mathrm{SR}}^2)\gamma_{\mathrm{arm}}}{1 + r_{\mathrm{SR}}^2 + 2r_{\mathrm{SR}}\cos(2\phi_{\mathrm{SR}})},$$
(3.16b)

where r_{SR} is the amplitude reflectivity of the signal-recycling mirror (SRM), and ϕ_{SR} is the phase detuning of the SR cavity. If the main homodyne detector (MHD) measures:

3.3 Quantum Noise Calculation

$$\hat{b}_\phi(\Omega) = \sqrt{\eta}[\sin\phi \hat{b}_A(\Omega) + \cos\phi \hat{b}_\varphi(\Omega)] + \sqrt{1-\eta}\hat{v}(\Omega), \qquad (3.17)$$

where ϕ is the homodyne angle and $\hat{v}$ is the additional vacuum state due to the non-unity quantum efficiency of the photodiode, and then the corresponding h-referred quantum-noise spectral density can be written as:

$$S_h(\Omega) = h_{\text{SQL}}^2 \frac{(\sin\phi\cos\phi)\widetilde{C}\widetilde{S_I}\widetilde{C}^\dagger(\sin\phi\cos\phi)^\text{T} + \frac{1-\eta}{\eta}|M|^2}{(\sin\phi\cos\phi)\widetilde{D}\widetilde{D}^\dagger(\sin\phi\cos\phi)^\text{T}}. \qquad (3.18)$$

This can be minimized by adjusting the squeezing angle θ of $\hat{s}$ and $\hat{p}$. We can estimate the optimal θ qualitatively from the asymptotic behavior of the resulting noise spectrum. At very high frequencies ($\Omega \gg \gamma$), from Eq. (3.2), $\hat{a} \sim \hat{s}$ and thus:

$$S_h(\Omega) \propto \cosh(2r_s) + \cos[2(\phi+\theta_s)]\sinh(2r_s). \qquad (3.19)$$

If the squeezing angle of $\hat{s}$:

$$\theta_s = \frac{\pi}{2} + n\pi - \phi, \qquad (3.20)$$

where n is integer, we achieve the optimal case, namely $S_h \propto e^{-2r_s}$. Similarly, at very low frequencies ($\Omega \ll \gamma$), we have $\hat{a} \sim \hat{p}$ and the spectral density $S_h \propto e^{-2r_p}$ if

$$\theta_p = \arctan\left[\frac{2\cos\phi\sin\beta}{\cos(\beta-\phi)+3\cos(\beta+\phi)}\right]. \qquad (3.21)$$

More accurate values for optimal $\theta_{s,p}$ can be obtained numerically, as we will show in the next section. Given optimal $\theta_{s,p}$, the sensitivity of this double squeezed-input scheme improves at both high and low frequencies. However, for the same reason as in the case of single squeezed-input that the two outputs of the filter cavity are entangled [9], this double squeezed-input scheme does not perform well in the intermediate frequency range. To recover the sensitivity, we need to use an additional homodyne detector (AHD) at the idle port E of the filter cavity. The corresponding measured quantity is:

$$\hat{q}_\zeta(\Omega) = \sqrt{\eta}[\sin\zeta \hat{b}_A(\Omega) + \cos\zeta \hat{b}_\varphi(\Omega)] + \sqrt{1-\eta}\hat{v}'(\Omega) \qquad (3.22)$$

where ζ is the homodyne angle and $v'(\Omega)$ is the additional vacuum state, which enters due to the non-unity quantum efficiency of photodiode. We combine $\hat{q}_\zeta(\Omega)$ with the output $\hat{b}_\phi(\Omega)$ using a linear filter $\mathcal{K}(\Omega)$, obtaining:

$$\hat{o}(\Omega) = \hat{b}_\phi(\Omega) - \mathcal{K}(\Omega)\hat{q}_\zeta(\Omega). \qquad (3.23)$$

Correspondingly, the noise spectral density of this new output $\hat{o}(\Omega)$ can be written as:

$$S_{\hat{o}}(\Omega) = S_{\hat{b}_\phi}(\Omega) - 2\Re[\mathcal{K}(\Omega) S_{\hat{b}_\phi,\hat{q}_\zeta}(\Omega)] + |\mathcal{K}(\Omega)|^2 S_{\hat{q}_\zeta}(\Omega). \tag{3.24}$$

The minimum quantum noise is achieved when $\mathcal{K}(\Omega) = S_{\hat{b}_\phi,\hat{q}_\zeta}(\Omega)/S_{\hat{b}_\phi}(\Omega)$, and theresulting h-referred noise spectrum with the AHD is then:

$$S_h^{\text{AHD}}(\Omega) = S_h(\Omega) - h_{\text{SQL}}^2 \frac{\eta|(\sin\phi\cos\phi)\widetilde{C}\widetilde{S}_{\text{IE}}(\sin\zeta\cos\zeta)^{\text{T}}|^2}{(\sin\phi\cos\phi)\widetilde{D}\widetilde{D}^\dagger(\sin\phi\cos\phi)^{\text{T}} S_\zeta(\Omega)}, \tag{3.25}$$

where $S_\zeta(\Omega) \equiv \eta(\sin\zeta\cos\zeta)\widetilde{S}_{\text{E}}(\sin\zeta\cos\zeta)^{\text{T}} + 1 - \eta$. The second term has a minus sign, which shows explicitly that the sensitivity increases as a result of the additional detection.

3.4 Numerical Optimizations

In this section, we will take into account realistic technical noise, and numerically optimize interferometer parameters for detecting GW signals from specific astrophysics sources, which include Neutron-Star–Neutron-Star (NS–NS) binaries and Bursts.

For a binary system, according to Ref. [13], spectral density of the GW signal is given by

$$S_h(2\pi f) = \frac{\pi}{12} \frac{(G\mathcal{M})^{5/3}}{c^3 r^2} \frac{\Theta(f_{\max} - f)}{(\pi f)^{7/3}}. \tag{3.26}$$

Here the "chirp" mass $\mathcal{M}$ is defined as $\mathcal{M} \equiv \mu^{3/5} M^{2/5}$ with μ and M being the reduced mass and total mass of the binary system. With other parameters being fixed, the corresponding spectrum shows a frequency dependence of $f^{-7/3}$. Therefore, as a measure of the detector sensitivity, we can define an integrated signal-to-noise ratio (SNR) for NS–NS binaries as:

$$\rho_{\text{NSNS}}^2 \propto \int_{f_{\min}}^{f_{\max}} \frac{f^{-7/3} df}{S_h^{\text{quant}}(2\pi f) + S_h^{\text{tech}}(2\pi f)}. \tag{3.27}$$

The upper limit of the integral $f_{\max} \sim f_{\text{ISCO}} \approx 4400 \times (M/M_\odot)$ Hz is determined by the innermost stable circular orbit (ISCO) frequency, and the lower limit $f_{\min}$ is set to be 10 Hz, at which the noise can no longer be considered as stationary. Here we choose $M = 2.8 M_\odot$, which is the same as in Ref. [8]. Here S_h^{quant} is the quantum noise spectrum derived in the previous sections and S_h^{tech} corresponds to the technical noise obtained from Bench [22].

Other interesting astrophysical sources are Bursts [32]. The exact spectrum is not well-modeled and a usual applied simple model is to assume a logarithmic-flat signal spectrum, i.e. $S_h(2\pi f) \propto f^{-1}$. The corresponding integrated SNR is then given by:

3.4 Numerical Optimizations

Table 3.2 Optimization results for NS–NS SNRs

Configurations	Parameters							
	ρ	r_{SR}	ϕ_{SR}	γ_1	γ_E	ϕ	θ_p	ζ
AdvLIGO	1.0	0.8	1.4	—	—	−1.0	—	—
FISAdvLIGO	1.0	0.7	1.5	—	—	−1.2	−0.6[§]	—
CMW	1.0	0.8	1.6	240	240	−0.8	−0.6	—
CMWA	1.2	0.7	1.6	230	210	0.0	−0.1	0.7

[§] This is the squeezing angle θ in the case of single squeezed-input.

Table 3.3 Optimization results for Bursts SNRs

Configurations	Parameters							
	ρ	r_{SR}	ϕ_{SR}	γ_1	γ_E	ϕ	θ_p	ζ
AdvLIGO	1.0	0.7	1.5	—	—	−0.2	—	—
FISAdvLIGO	1.5	0.8	1.6	—	—	0.0	−1.6[§]	—
CMW	1.5	0.8	1.6	0.0	0.0	0.0	−1.4	—
CMWA	1.5	0.8	1.6	140	140	0.0	−0.2	0.9

$$\rho^2_{\text{Bursts}} \propto \int_{f_{\min}}^{f_{\max}} \frac{d\log f}{S_h^{\text{quant}}(2\pi f) + S_h^{\text{tech}}(2\pi f)}. \quad (3.28)$$

The integration limit is taken to be the same as in the NS–NS binaries case.

To estimate the SNR and also motivate future implementation of this scheme, we assume the filter cavity has a length of ~ 30 m and an achievable optical loss of 10 ppm per bounce and also consider the non-unity quantum efficiency of the photodiodes $\eta = 0.9$ for both MHD and AHD. Other relevant parameters will be further optimized numerically. For comparison, we will also optimize other related configurations, which includes AdvLIGO, AdvLIGO with frequency-independent squeezed-input (FISAdvLIGO for short), and the CMW scheme. Specifically, the free parameters for these different schemes that need to be optimized are the following:

$$\text{AdvLIGO:} \quad r_{SR}, \phi_{SR}, \phi, \quad (3.29a)$$

$$\text{FISAdvLIGO:} \quad r_{SR}, \phi_{SR}, \phi, \theta, \quad (3.29b)$$

$$\text{CMW:} \quad r_{SR}, \phi_{SR}, \phi, \gamma_1, \gamma_E, \theta_s, \theta_p, \quad (3.29c)$$

$$\text{CMWA:} \quad r_{SR}, \phi_{SR}, \phi, \gamma_1, \gamma_E, \theta_s, \theta_p, \zeta. \quad (3.29d)$$

The resulting optimal parameters for different schemes are listed in Tables 3.2 and 3.3. They are rounded to have two significant digits at most in view of various uncertainties in the technical noise. The integrated SNR ρ is normalized with respect to that of the AdvLIGO configuration. The optimal θ_s from the numerical result is in accord with the asymptotic estimation, namely $\theta_s \approx (\pi/2) - \phi$ (cf. Eq. (3.20)).

The corresponding quantum-noise spectrums of different schemes optimized for detecting gravitational waves from NSNS binaries are shown in Fig. 3.2. The

Fig. 3.2 Quantum-noise spectra of different schemes with optimized parameters for detecting gravitational waves from NS–NS binaries. The optimal values for the parameters are listed in Table. 3.2. From Ref. [24]

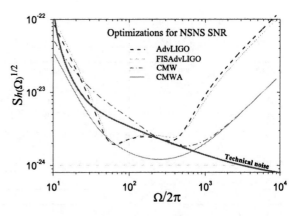

Fig. 3.3 Quantum-noise spectra of different schemes which are optimized for detecting GWs from Bursts. The optimal values for the parameters are listed in Table. 3.3. From Ref. [24]

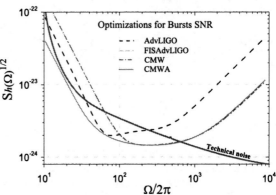

AdvLIGO, FISAdvLIGO and CMW schemes almost have the identical integrated sensitivities and the CMWA scheme shows a moderate 20% improvement in SNR. This is attributable to the fact that the signal spectrum of NSNS binaries has a $f^{-7/3}$ dependence and low-frequency sensitivity is very crucial. However, due to low-frequency technical noise, advantages of the CMWA scheme are at most limited in the case for detecting low-frequency sources.

The case for detecting GWs from Bursts is shown in Fig. 3.3. All other three schemes have a significant 50% improvement in terms of SNR over AdvLIGO. The sensitivities of the optimal FISAdvLIGO, CMW and CMWA at high frequencies almost overlap each other. In addition, the detuned phase ϕ_{SR} of the signal-recycling cavity of those three are nearly equal to $\pi/2$, which significantly increases the effective detection bandwidth of the gravitational-wave detectors and is the same as in the Resonant-Sideband Extraction (RSE) scheme. This is because a broadband sensitivity is preferable in the case of Bursts which have a logarithmic-flat spectrum.

To show explicitly how different parameters affect the sensitivity of the CMWA scheme, we present the quantum-noise spectra of different schemes, by using the same parameters as the optimal CMWA in Fig. 3.4. In the case of AdvLIGO, we obtain

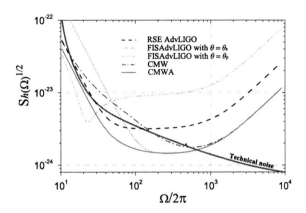

Fig. 3.4 Quantum-noise spectrums of different schemes using the same parameters as the optimal CMWA to show how different parameters affect sensitivity of the CMWA scheme. From Ref. [24]

a RSE configuration with $\phi_{SR} \approx \pi/2$. The quantum noise of FISAdvLIGO with squeezing angle $\theta = \theta_s$ is lower at high frequencies, but higher at low frequencies than the RSE AdvLIGO. FISAdvLIGO with $\theta = \theta_p$ behaves in the opposite way, with a significant increase of sensitivity at low frequencies but worse sensitivity at high frequencies. The CMW scheme with double squeezed-input, just as expected, can improve the sensitivity at both high and low frequencies but performs not so well at intermediate frequencies. The CMWA scheme performs very nicely over the whole observational band compared with others and it would be more attractive if the technical noise of the AdvLIGO design could be further decreased.

3.5 Conclusions

We have proposed and analyzed the double squeezed-input CMWA scheme as an option for increasing the sensitivity of future advanced GW detectors. Given an achievable optical loss of the filter cavity, and 10 dB squeezing, this CMWA configuration shows a noticeable reduction in quantum noise at both high and low frequencies compared with other schemes. Since the length of the filter cavity considered here is around 30 m, with the development of better low-loss coating and squeezing-state sources, the CMWA scheme could be a promising and relatively simple add-on to AdvLIGO, without needing to dramatically modify the existing interferometer topology.

References

1. O. Arcizet, T. Briant, A. Heidmann, M. Pinard, Beating quantum limits in an optomechanical sensor by cavity detuning. Phys. Rev. A **73**, 033819 (2006)
2. V.B. Braginsky, Classical and quantum restrictions on the detection of weak disturbances of a macroscopic oscillator. JETP **26**, 831 (1968)
3. V.B. Braginsky, F.Y. Khalili, *Quantum Measurement* (Cambridge University Press, Cambridge, 1992)

4. V.B. Braginsky, F.Y. Khalili, Low noise rigidity in quantum measurements. Phys. Lett. A **257**, 241–246 (1999)
5. V.B. Braginsky, M.L. Gorodetsky, F.Y. Khalili, K.S. Thorne, Noise in gravitational-wave detectors and other classical-force measurements is not influenced by test-mass quantization. Phys. Rev. D **61**, 044002 (2000)
6. A. Buonanno, Y. Chen, Quantum noise in second generation, signalrecycled laser interferometric gravitational-wave detectors. Phys. Rev. D **64**, 042006 (2001)
7. A. Buonanno, Y. Chen, Signal recycled laser-interferometer gravitational-wave detectors as optical springs. Phys. Rev. D **65**, 042001 (2002)
8. A. Buonanno, Y. Chen, Scaling law in signal recycled laserinterferometer gravitational-wave detectors. Phys. Rev. D **67**, 062002 (2003)
9. A. Buonanno, Y. Chen, Improving the sensitivity to gravitational-wave sources by modifying the input-output optics of advanced interferometers. Phys. Rev. D **69**, 102004 (2004)
10. Y. Chen, Sagnac interferometer as a speed-meter-type, quantumnondemolition gravitational-wave detector. Phys. Rev. D **67**, 122004 (2003)
11. LIGO Scientific Collaboration, First upper limits from LIGO on gravitational wave bursts. Phys. Rev. D **69**, 102001 (2004)
12. T. Corbitt, N. Mavalvala, S. Whitcomb, Optical cavities as amplitude filters for squeezed fields. Phys. Rev. D **70**, 022002 (2004)
13. S.L. Danilishin, Sensitivity limitations in optical speed meter topology of gravitational-wave antennas. Phys. Rev. D **69**, 102003 (2004)
14. F. Khalili, Limiting Sensitivity During the Continuous Tracking of the Coordinates of a Quantum Test System, vol. 294 (Akademiia Nauk SSSR, Doklady, 1987), p. 603
15. http://geo600.aei.mpg.de.
16. http://ilog.ligo wa.caltech.edu:7285/advligo/Bench.
17. http://tamago.mtk.nao.ac.jp.
18. http://www.ligo.caltech.edu.
19. http://www.ligo.caltech.edu/advLIGO.
20. http://www.virgo.infn.it.
21. M.T. Jaekel, S. Reynaud, Quantum limits in interferometric measurements. EPL (Europhys. Lett.) **13**(4), 301 (1990)
22. F.Y. Khalili, Frequency-dependent rigidity in large-scale interferometric gravitational-wave detectors. Phys. Lett. A **288**, 251–256 (2001)
23. F.Y. Khalili, Increasing future gravitational-wave detectors' sensitivity by means of amplitude filter cavities and quantum entanglement. Phys. Rev. D **77**, 062003 (2008)
24. F.Y. Khalili, Y. Levin, Speed meter as a quantum nondemolition measuring device for force. Phys. Rev. D **54**, 004735 (1996)
25. F.Y. Khalili, V.I. Lazebny, S.P. Vyatchanin, Sub-standard-quantumlimit sensitivity via optical rigidity in the advanced LIGO interferometer with optical losses. Phys. Rev. D **73**, 062002 (2006)
26. F.Y. Khalili, H. Miao, Y. Chen, Increasing the sensitivity of future gravitational-wave detectors with double squeezed-input. Phys. Rev. D **80**, 042006 (2009)
27. H.J. Kimble, Y. Levin, A.B. Matsko, K.S. Thorne, S.P. Vyatchanin, Conversion of conventional gravitational-wave interferometers into quantum nondemolition interferometers by modifying their input and/or output optics. Phys. Rev. D **65**, 022002 (2001)
28. I.S. Kondrashov, D.A. Simakov, F.Y. Khalili, S.L. Danilishin, Optimizing the regimes of the Advanced LIGO gravitational wave detector for multiple source types. Phys. Rev. D **78**, 062004 (2008)
29. K. McKenzie, N. Grosse, W.P. Bowen, S.E. Whitcomb, M.B. Gray, D.E. McClelland, P.K. Lam, Squeezing in the Audio Gravitational-Wave Detection Band. Phys. Rev. Lett **93**, 161105 (2004)
30. A.F. Pace, M.J. Collett, D.F. Walls, Quantum limits in interferometric detection of gravitational radiation. Phys. Rev. A **47**, 3173 (1993)

31. K.A. Postnov, L.R. Yungelson, The Evolution of Compact Binary Star Systems. Living. Rev. Relativ. **9**, 6 (2006)
32. P. Purdue, Analysis of a quantum nondemolition speed-meter interferometer. Phys. Rev. D **66**, 022001 (2002)
33. P. Purdue, Y. Chen, Practical speed meter designs for quantum nondemolition gravitational-wave interferometers. Phys. Rev. D **66**, 122004 (2002)
34. K.S. Thorne, *Three Hundred Years of Gravitation* (Cambridge University Press, Cambridge, 1987)
35. E.G. Unruh, *Quantum Optics Experimental Gravitation and Measurement Theory* (Springer-Verlag, Berlin, 1982)
36. H. Vahlbruch, S. Chelkowski, B. Hage, A. Franzen, K. Danzmann, R. Schnabel, Coherent Control of Vacuum Squeezing in the Gravitational- Wave Detection Band. Phys. Rev. Lett **97**, 011101 (2006)
37. S.P. Vyatchanin, A.B. Matsko, Quantum limit on force measurements. JETP **77**, 218 (1993)

Chapter 4
Modifying Test-Mass Dynamics: Double Optical Spring

4.1 Preface

In this chapter, we will discuss the approach to surpassing the free-mass Standard Quantum Limit (SQL) by modifying the the test-mass dynamics with double optical springs. We explore the frequency dependence of the optical spring effect. In particular, we show that the frequency dependence of double optical springs allows us to create a "negative inertia", which cancels the positive inertia of the test-mass with the mechanical response significantly enhanced. This can surpass the free-mass SQL over a broad frequency range. In addition, we show the feasibility of demonstrating such an effect with Gingin high optical power test facility. The same setup could eventually be implemented in future advanced GW detectors. This is a joint effort by Farid Khalili, Stefan Danilishin, Helge Mueller-Ebhardt, Yanbei Chen, Chunnong Zhao, and myself. It is an ongoing research project.

4.2 Introduction

There are two types of fundamental quantum noise in an advanced gravitational-wave (GW) detector: one, known as measurement shot noise, originates from fundamental phase fluctuations of light waves, while the other one, caused by quantum fluctuations of light wave amplitude, results in fluctuations of radiation pressure on the test-masses of the interferometer, called the back-action noise. If these two are not correlated, they impose a lower bound on the detector sensitivity, which is in essence the *Standard Quantum Limit* (SQL) [1, 5, 2]. For a free probe mass, the SQL, written in terms of GW strain amplitude spectrum, is given by:

$$h_{\text{SQL}}^{\text{fm}} = \sqrt{\frac{2\hbar}{m\Omega^2 L^2}}, \tag{4.1}$$

where m is the reduced mass, and L the arm length of an interferometric GW detector. In order to further improve the sensitivity, many approaches were proposed in the literature to surpass this limit. Most of them, such as variational measurement [7], are based on the utilization of the cross correlations between the shot noise and back-action noise. These methods, however, are extremely susceptible to the influence of optical loss, and impose the following severe limitation on the amount by which these methods can beat the SQL (see, e.g., Ref. [6]):

$$\xi = \frac{h}{h_{\text{SQL}}^{\text{fm}}} = \sqrt[4]{\frac{1-\eta}{\eta}}. \tag{4.2}$$

Here, h is the sensitivity of the GW detector in terms of the GW strain spectrum; η is the normalized quantum efficiency of the meter readout. Estimates show that, at present, η is limited mostly by the photodetector quantum efficiency. Given the values of η span a range of 0.95–0.99 (considered as moderately optimistic), this yields:

$$\xi \gtrsim 0.5 - 0.3. \tag{4.3}$$

which is not very useful.

There is another method to overcome the free-mass SQL, which is not influenced by optical loss, and thus not susceptible to this limitation. Instead of reducing the total noise, it increases the signal displacement of the test mass, and thus surpasses the free-mass SQL by converting an initially free test-mass into a more "responsive" object, e.g., a harmonic oscillator. In the general case, the SQL for force detection is equal to:

$$\sqrt{S_{\text{SQL}}^{F}(\Omega)} = \sqrt{2\hbar m |\chi^{-1}(\Omega)|}, \tag{4.4}$$

where χ/m is the mechanical susceptibility, which is in essence the system Green's function Fourier transform. Correspondingly, the SQL for the system characterized by χ in terms of GW strain is equal to

$$h_{\text{SQL}} = \frac{1}{L}\frac{\sqrt{S_{\text{SQL}}^{F}(\Omega)}}{m\Omega^2} = h_{\text{SQL}}^{\text{fm}}\sqrt{\frac{|\chi^{-1}(\Omega)|}{\Omega^2}} \tag{4.5}$$

For a free mass, $\chi^{-1} = -\Omega^2$, and the factor in the square root is simply unity, while for a mechanical oscillator $\chi^{-1} = -(\Omega^2 - \omega_m^2) + i\gamma_m\Omega$, and the corresponding SQL is smaller than the one for the free mass by a significant factor of $\sqrt{\omega_m/\gamma_m}$, which can be 10^4 for a high-Q oscillator.

To modify the dynamics of a free test mass into that of a harmonic oscillator, a well-known approach of using optical rigidity induced by the position-dependent radiation-pressure force is widely used. Unfortunately, with a single optical spring, only narrow-band gain can be obtained. We show here that by using two optical

4.2 Introduction

springs and exploiting their frequency dependence, it is possible to create a *negative optical inertia* [9], which compensates positive inertia of the test mass, and thus allows the free-mass SQL to be surpassed over a broad frequency band.

The outline of this chapter is as follows: in Sect. 4.3, we give a general mathematical treatment of the optical negative inertial and derive the resulting mechanical response function; in Sect. 4.4, we try to address the problem when the cavity bandwidth is an issue as in case of Gingin setup; in Sect. 4.6, we conclude our results and outline future works in this direction.

4.3 General Considerations

It is shown in Ref. [4] that a signal-recycled interferometric GW detector can be mapped into a detuned Fabry-Pérot optical cavity. We do not need to go into the details of an interferometer and can only consider a single cavity for a general discussion. As previously proposed in Ref. [10], instead of a single carrier light beam considered in Ref. [4], here we include two carrier light beams with different frequencies, and each of them will induce an optical rigidity and affect the dynamics of the test mass. The resulting effective mechanical susceptibility χ of the test mass can be read off from the response of the test mass displacement x to the external force F, namely:

$$x(s) = \frac{\chi(s)F(s)}{m} \implies \chi^{-1}(s) = s^2 + \frac{K_1(s) + K_2(s)}{m}, \quad (4.6)$$

with $s = -i\Omega$. The optical rigidities $K_{1,2}$ are given by:

$$K_{1,2} = \frac{mJ_{1,2}\delta_{1,2}}{s^2 + 2\gamma s + \Delta_{1,2}^2} \quad (4.7)$$

with

$$J_{1,2} = \frac{4\omega_{1,2}I_{1,2}}{mcL}, \quad \Delta_{1,2} = \sqrt{\gamma^2 + \delta_{1,2}}. \quad (4.8)$$

Here, γ is the cavity bandwidth, which is equal to $\gamma = \frac{\pi c}{2L\mathcal{F}}$, with $\mathcal{F}$ the optical finesse; $\omega_{1,2}$ are the frequencies of two carrier light beams; $I_{1,2}$ are the intracavity powers; $\delta_{1,2} = \omega_{c1,2} - \omega_0$ are the detunings with respect to the cavity resonant frequency given by ω_0.

The negative inertia regime is reached when:

$$K_1(0) + K_2(0) = 0, \quad \frac{1}{2}\frac{\partial^2[K_1(s) + K_2(s)]}{\partial s^2}\bigg|_{s=0} + m = 0. \quad (4.9)$$

The first equation indicates that the static rigidity from two carrier light beams cancel each other while the second equation gives a zero inertia which can significantly enhance the mechanical response to the external force (i.e., the GW tidal force, in the context here). These two conditions provide us with the following two equations:

$$\frac{J_1\delta_1}{\Delta_1^2} + \frac{J_2\delta_2}{\Delta_2^2} = 0, \qquad \frac{J_1\delta_1(\Delta_1^2 - 4\gamma^2)}{\Delta_1^6} + \frac{J_2\delta_2(\Delta_2^2 - 4\gamma^2)}{\Delta_2^6} = 1. \qquad (4.10)$$

Solving these equations with respect to $J_1\delta_1$ and $J_2\delta_2$, one gets:

$$J_1\delta_1 = \frac{\Delta_1^2}{\left(\frac{1}{\Delta_1^2} - \frac{1}{\Delta_2^2}\right)\left(1 - \frac{4\gamma^2}{\Delta^2}\right)}, \quad J_2\delta_2 = \frac{\Delta_2^2}{\left(\frac{1}{\Delta_2^2} - \frac{1}{\Delta_1^2}\right)\left(1 - \frac{4\gamma^2}{\Delta^2}\right)}, \qquad (4.11)$$

where

$$\Delta^2 = \left(\frac{1}{\Delta_1^2} + \frac{1}{\Delta_2^2}\right)^{-1}. \qquad (4.12)$$

In the practically interesting case of

$$\delta_1 \sim \delta_2 \gg \gamma, \qquad (4.13)$$

we have:

$$J_1 = \frac{\delta_1}{\left(\frac{1}{\delta_1^2} - \frac{1}{\delta_2^2}\right)}, \quad J_2 = \frac{\delta_2}{\left(\frac{1}{\delta_2^2} - \frac{1}{\delta_1^2}\right)}, \qquad (4.14)$$

which implies the following relation between the optical powers of the two carriers and their detunings:

$$\frac{J_1}{J_2} \equiv \frac{I_{c1}}{I_{c2}} \approx -\frac{\delta_1}{\delta_2}. \qquad (4.15)$$

One can easily see that these detunings should have opposite signs, in order to compensate for the static rigidity. Note also that because $J_{1,2} > 0$, the detuning with the larger magnitude has to be negative.

Substituting Eq. (4.11) into Eq. (4.6), one gets the following expression for the mechanical response function:

$$\chi^{-1}(s) = s^2 - \frac{\Delta_1^2\Delta_2^2(s^2 + 2\gamma s)}{(s^2 + 2\gamma s + \Delta_1^2)(s^2 + 2\gamma s + \Delta_2^2)\left(1 - 4\gamma^2/\Delta^2\right)}. \qquad (4.16)$$

In order to reveal the characteristic features of this expression, we expand it into a Taylor series in s:

4.3 General Considerations

$$\chi^{-1}(s) \approx \frac{1}{1-\frac{4\gamma^2}{\Delta^2}} \left[-2\gamma s + 4\gamma s^3 \left(\frac{1}{\Delta^2} - 2\gamma^2 \frac{\Delta_1^4 + \Delta_1^2\Delta_2^2 + \Delta_2^4}{\Delta_1^4 \Delta_2^2} \right) \right.$$
$$\left. + s^4 \left(\frac{1}{\Delta^2} - 12\gamma^2 \frac{\Delta_1^4 + \Delta_1^2\Delta_2^2 + \Delta_2^4}{\Delta_1^4 \Delta_2^2} + 16\gamma^4 \frac{\Delta_1^4 + \Delta_2^4}{\Delta^2 \Delta_1^4 \Delta_2^4} \right) \right] + O(s^5). \tag{4.17}$$

Assuming again that $\delta_1 \sim \delta_2 \gg \gamma$, we obtain the following simple expression for the mechanical response function:

$$\chi^{-1}(s) \approx -2\gamma s + \frac{4\gamma s^3}{\Delta^2} + \frac{s^4}{\Delta^4} + O(s^5). \tag{4.18}$$

4.4 Further Considerations: Removing the Friction Term

In Eq. (4.18) of the previous section, there is a friction term $\sim \gamma s$ in the effective response function, which limits the low-frequency improvement. It will not be important if the cavity bandwidth is narrow, as mentioned. However, to experimentally demonstrate the "negative inertia" effect with the Gingin High Power Facility, it would be rather challenging for the given specifications. One might think that this friction term could be removed by using a feedback control. However, this will not work because any classical feedback does not allow us to improve the detector sensitivity. Even though the controlled dynamics of the test mass will not have the friction term, the quantum limit will still be given by the original response function $\chi(\Omega)$ rather than the controlled one. This is shown explicitly in Ref. [3], where a feedback control is used to control the instability induced by the optical spring, while the sensitivity for detecting GWs remains unaffected.

We thus need to remove the friction term internally with the help of the optical spring (a quantum feedback control method) instead of externally with a classical feedback control. For this purpose, in addition to Eq. (4.9), we need to further impose the following requirement on the optical spring:

$$\frac{\partial [K_1(s) + K_2(s)]}{\partial s}\bigg|_{s=0} = 0. \tag{4.19}$$

It seems rather trivial to satisfy this condition, as we have five free parameters: $J_{1,2}, \delta_{1,2}$ and γ. As it turns out, there are no reasonable solutions (imaginary roots). To solve this problem, we realize that in an actual signal-recycled interferometer, the signal-recycling cavity acts as an effective mirror and one can tune cavity bandwidth of two carriers over a large range [4, 8]. In the case of the Gingin setup, a three-mirror coupled cavity can achieve the same effect. Therefore, we do not need to assume the same cavity bandwidth γ for both carriers. Given different cavity bandwidths $\gamma_{1,2}$, we find that sthe following achievable specifications can satisfy both conditions in Eqs. (4.9) and (4.19):

$$\Delta_1/2\pi = 200\,\text{Hz}, \Delta_2/2\pi = -500\,\text{Hz}, \gamma_1/2\pi = 36\,\text{Hz}, \gamma_2/2\pi = 400\,\text{Hz}. \tag{4.20}$$

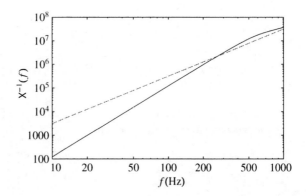

Fig. 4.1 The resulting response function with a double optical spring (*solid*), compared with that of a free mass (*dashed*)

With test mass $m = 0.8\,\mathrm{kg}$ and a cavity length $L = 80\,\mathrm{m}$, the intracavity powers for the two carriers are

$$I_1 = 3\,\mathrm{kW},\ I_2 = 10\,\mathrm{kW}. \tag{4.21}$$

The resulting response function, compared with that of the free-mass, is shown in the Fig. 4.1. As we can see, this setup provides us with significant improvement at low frequencies.

4.5 "Speed-Meter" Type of Response

In this section, we will look at another parameter regime, in which two optical springs have opposite detunings, and they cancel each other. In this case, the test mass dynamics is not modified, but we can achieve a "speed-meter" type of response. Unlike a typical speed meter, which allows us to surpass the SQL at low frequencies, the scheme considered does not. However, its sensitivity can follow the SQL over a large range at low frequency, which is also very interesting.

The "speed-meter" type of response can be achieved if we choose the right quadrature. From Eqs. (47) and (48) in the scaling-law paper [4], the response of the two output quadratures to the test-mass displacement is given by:

$$R_{Y_1 F} = \sqrt{\frac{\gamma \mathcal{I}_c}{2\hbar}} \frac{\Delta}{(\Omega - \Delta + i\gamma)(\Omega + \Delta + i\gamma)} \tag{4.22}$$

$$R_{Y_2 F} = -\sqrt{\frac{\gamma \mathcal{I}_c}{2\hbar}} \frac{(\gamma - i\Omega)}{(\Omega - \Delta + i\gamma)(\Omega + \Delta + i\gamma)} \tag{4.23}$$

where γ is the cavity bandwidth, and Δ is the cavity detuning, and $\mathcal{I}_c = 8\omega_0 I_c/(Lc)$. If we read out

$$\hat{Y}_\zeta = \hat{Y}_1 \sin\zeta + \hat{Y}_2 \cos\zeta \tag{4.24}$$

4.5 "Speed-Meter" Type of Response

at the quadrature angle

$$\zeta = \arctan\frac{\gamma}{\Delta}, \qquad (4.25)$$

the resulting response of $\hat{Y}_\zeta$ to the displacement will be:

$$R_{Y_\zeta F} = \sqrt{\frac{\gamma \mathcal{I}_c}{2\hbar}} \frac{i\Omega\cos\zeta}{(\Omega - \Delta + i\gamma)(\Omega + \Delta + i\gamma)}. \qquad (4.26)$$

For a small cavity bandwidth $\gamma < \Delta$, and at low frequencies $\Omega < \Delta$, so we simply have

$$R_{Y_\zeta F} \approx \sqrt{\frac{\gamma \mathcal{I}_c}{2\hbar}} \frac{\cos\zeta}{\Delta^2}(-i\Omega). \qquad (4.27)$$

This basically indicates that the detector has a "speed-meter" type of response.

In our case, we consider that the two carriers A and B have opposite detunings so that the optical-spring effects perfectly cancel each other, as discussed in the double-optical-spring (DOS) paper [10]. If we further require the detection quadrature angle $\zeta_A = -\zeta_B$ is opposite for two carriers, the responses of them [(cf. Eq. (4.27))] will be the same, $\hat{Y}_\zeta^A = \hat{Y}_\zeta^B$, and we essentially have two identical probes. By combining their two readouts with kernel functions K_A and K_B, we have:

$$\hat{Y}_{\text{tot}} = K_A(\Omega)\hat{Y}_A + K_B(\Omega)\hat{Y}_B. \qquad (4.28)$$

By optimizing K_A and K_B using the technique introduced in the Ref. [10], we can obtain maximal sensitivity. For a numerical estimate, we choose the following specifications:

$$\Delta/(2\pi) = 100\,\text{Hz}, \gamma/(2\pi) = 50\,\text{Hz}, \zeta = \arctan(0.5), I_A = I_B = 400\,\text{kW}. \qquad (4.29)$$

The resulting noise curve and the optimal kernel functions are shown in Fig. 4.2. At low frequencies, we almost follow the SQL. There is actually an analytical formula for the noise curve, which is:

$$S^h(\Omega) = \frac{1}{L^2}\left[\frac{1}{|R_{Y_\zeta F}|^2} + \frac{S_{\text{SQL}}^2(\Omega)}{4}|R_{Y_\zeta F}|^2\right] \geq S_{\text{SQL}}^h, \qquad (4.30)$$

where $S_{\text{SQL}} = \hbar/(m\Omega^2)$ and the optical power is twice that of an individual carrier. Given the response in (4.27), we have:

$$S^h(\Omega) = S_{\text{SQL}}^h\left[\frac{1}{|R_{Y_\zeta F}|^2 S_{\text{SQL}}} + \frac{S_{\text{SQL}}(\Omega)}{4}|R_{Y_\zeta F}|^2\right] \propto S_{\text{SQL}}^h. \qquad (4.31)$$

Thiss basically means that we can follow the SQL but never be able to surpass it. This is in contrast with a true "speed meter" in which the SQL can be surpassed significant as long as we have a sufficient large optical power. The next question would be how we could recover the true "speed-meter" response in the DOS configuration.

Fig. 4.2 The *top* figure shows the DOS noise curve normalized with respect to the standard quantum limit at 100 Hz. The *bottom* one shows the corresponding optimal kernel functions for the two carriers

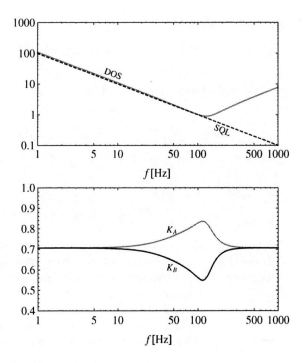

4.6 Conclusions and Future Work

We have shown the possibility of a "negative inertia" effect attributable to the specific frequency dependence of double optical springs. This is capable of reducing the effective inertia of the probe mass, and thus can significantly enhance the mechanical response of the interferometer to GW signals. This effect allows the interferometer to surpass the free-mass SQL over a wide frequency band. There are several issues that need to be further investigated: (1) the stability of the system. What feedback control method is necessary to stabilize the system? (cf. the discussions in Ref. [10] for the single optical spring case); (2) if one uses a third carrier for readout purposes, how much improvement in sensitivity could be obtained, compared to the current double-carrier case? In addition, we have considered the "speed-meter" type of response, and have shown that double optical springs allow us to follow the SQL at low frequencies.

References

1. V.B. Braginsky, Classical and quantum restrictions on the detection of weak disturbances of a macroscopic oscillator. JETP **26**, 831 (1968)
2. V.B. Braginsky, F.Y. Khalili, *Quantum Measurement* (Cambridge University Press, Cambridge, 1992)
3. A. Buonanno, Y. Chen, Signal recycled laser-interferometer gravitational-wave detectors as optical springs. Phys. Rev. D **65**, 042001 (2002)

References

4. A. Buonanno, Y. Chen, Scaling law in signal recycled laserinterferometer gravitational-wave detectors. Phys. Rev. D **67**, 062002 (2003)
5. C.M. Caves, K.S. Thorne, R.W. Drever, V.D. Sandberg, M. Zimmermann, On the measurement of a weak classical force coupled to a quantummechanical oscillator. I. Issues of principle. Rev. Mod. Phys. **52**, 341 (1980)
6. Y. Chen, S.L. Danilishin, F.Y. Khalili, H. Mller-Ebhardt, QND measurements for future gravitational-wave detectors, arXiv:0910.0319, 2009
7. H.J. Kimble, Y. Levin, A.B. Matsko, K.S. Thorne, S.P. Vyatchanin, Conversion of conventional gravitational-wave interferometers into quantum nondemolition interferometers by modifying their input and/or output optics. Phys. Rev. D **65**, 022002 (2001)
8. J. Mizuno, Comparison of optical configurations for laser-interferometric gravitational-wave detectors. Ph.D. thesis, Max-Planck Institut für Quantenoptik, Garching, Germany, 1995
9. H. Mueller-Ebhardt, On quantum effects in the dynamics of macroscopic test masses. Ph.D. thesis, Leibniz University Hannover, 2009
10. H. Rehbein, H. Müller-Ebhardt, K. Somiya, S.L. Danilishin, R. Schnabel, K. Danzmann, Y. Chen, Double optical spring enhancement for gravitational-wave detectors. Phys. Rev. D **78**, 062003 (2008)

Chapter 5
Measuring a Conserved Quantity: Variational Quadrature Readout

5.1 Preface

In this chapter, we consider surpassing the Standard Quantum Limit (SQL) for measuring a weak force with a mechanical oscillator. By using a time-dependent variational readout, we can measure the mechanical quadrature—a conserved quantity of the oscillator motion—and evade measurement-induced back-action. This is motivated by the pioneering work of Vyatchanin et al. [7, 8], in which such a back-action-evading variational scheme is proposed to detect a force signal with known arrival time. Here, we will go beyond such a limitation and make it suitable for all possible stationary signals, which can be characterized by their spectrum. This will be useful for: (i) improving the sensitivity of future GW detectors if the test-mass frequency (~ 1 Hz) is upshifted to the detection band (~ 100 Hz) by the optical-spring effect and (ii) improving the sensitivity of an atom force microscopy with a high-frequency mechanical oscillator as the probe. This is a continuing joint research effort by Stefan Danilishin, Yanbei Chen, and myself.

5.2 Introduction

Due to recent significant achievements in fabricating high-quality mechanical devices, electromechanical and optomechanical devices have played important roles in probing weak forces and tiny displacements. Two notable examples on the large scale and small scale are: (i) a gravitational-wave detector, in which kg-scale test-masses are coupled to high-power optical field for probing tiny ripples in the spacetime-gravitational waves [9], and (ii) an atomic-force microscope, in which a micromechanical oscillator coupled to an electric or optical field is used to probe atomic forces [6]. Their force sensitivity is limited by the SQL [1], and which has a spectral density of:

$$S_F^{\text{SQL}}(\Omega) = \frac{\hbar}{|R_{xx}(\Omega)|}, \tag{5.1}$$

where the mechanical response $R_{xx}(\Omega) = -1/m[(\Omega^2 - \omega_m^2)]$ for a high-Q oscillator with eigenfrequency ω_m. The origin of such a limit comes from the fact that we try to measure the external force by monitoring the changes in the oscillator position $\hat{x}$ at different times, which do not commute with each other, i.e.,

$$[\hat{x}(t), \hat{x}(t')] = -\frac{i\hbar}{m\omega_m} \sin \omega_m(t - t'). \tag{5.2}$$

This leads to the following Heisenberg uncertainty principle:

$$\Delta x(t)\Delta x(t') \geq \frac{\hbar}{2m\omega_m} \sin \omega_m(t - t'), \tag{5.3}$$

which basically means that the oscillator positions at different times cannot be simultaneously measured with an arbitrarily high accuracy; this gives rise to the SQL for the force sensitivity.

There are two approaches to overcoming such a limit [1]: (i) *a stroboscopic measurement*. Thiss measures the oscillator with pulses separated by an integer times the oscillation period. From the commutator relation Eq. (5.2), we learn that if $t - t' = 2n\pi/\omega_m$, $[\hat{x}(t), \hat{x}(t')] = 0$, and, then they are simultaneously measurable; (ii) *a quadrature measurement*. Instead of measuring the oscillator position, this measures quantum nondemolition (QND) observables-mechanical quadratures $\hat{X}_{1,2}$ which are defined through the equation:

$$\hat{x}(t) \equiv \hat{X}_1(t)\cos\omega_m t + \frac{\hat{X}_2(t)}{m\omega_m}\sin\omega_m t. \tag{5.4}$$

These observables satisfy

$$[\hat{X}_1(t), \hat{X}_1(t')] = [\hat{X}_2(t), \hat{X}_2(t')] = 0, \quad [\hat{X}_1(t), \hat{X}_2(t')] = i\hbar\delta(t-t')dt. \tag{5.5}$$

Therefore, if we are able to detect the force by measuring only one of the quadratures, e.g., $\hat{X}_1(t)$, there will not be an associated quantum limit. To realize such a measurement, Braginsky et al. proposed to modulate the electro/optomechanical interaction strength G at the mechanical frequency, namely $G(t) = G_0 \cos\omega_m t$ [2]. Specifically, the interaction Hamiltonian can be written as:

$$\hat{H}_{\text{int}} = \hbar G(t)\hat{x}(t)\hat{a}_1(t) = \hbar G_0 \cos\omega_m t \left[\hat{X}_1(t)\cos\omega_m t + \frac{\hat{X}_2(t)}{m\omega_m}\sin\omega_m t\right]\hat{a}_1(t)$$

$$= \frac{1}{2}\hbar G_0 \left[\hat{X}_1(t) + \hat{X}_1(t)\cos 2\omega_m t + \frac{\hat{X}_2(t)}{m\omega_m}\sin 2\omega_m t\right], \tag{5.6}$$

5.2 Introduction

where $\hat{a}_1(t)$ represents the electrical/optical degrees of freedom. If $\hat{a}_1(t)$ is changing slowly compared with $2\omega_m$ and it does not have a significant fluctuation at $2\omega_m$, we will measure the QND observable $\hat{X}_1$ without imposing any back-action. For optomechanical devices, this can be realized with a mechanical oscillator coupled to a high-finesse cavity, such that the cavity bandwidth γ is much smaller than the mechanical frequency, as proposed by Clerk et al. [3].

Here, we will consider another regime where the probing field is not slowly changing and has fluctuations at $2\omega_m$, e.g., one with a large cavity bandwidth in the case of a cavity-assisted scheme. With a time-domain variational readout, we can evade the measurement-induced back-action from the output data, thus realizing a quasi-QND measurement. This will be useful for small-scale atomic force microscopy, where a high-finesse cavity could be difficult to incorporate. It is also useful for large-scale GW detectors, because a large cavity bandwidth is preferred for increasing the detection bandwidth.

The outline of this chapter is as follows: In Sect. 5.3, we will analyze the dynamics of an optomechanical system. In Sect. 5.4, we will consider the time-domain variational scheme and demonstrate how the back action can be evaded if we effectively sense only one of the quadrature. One limitation for such a scheme is that it lose the resonant gain. In Sect. 5.5, we will introduce another variational scheme which mimics a stroboscopic measurement, in which we can recover the resonant gain by introducing an insignificant back-action.

5.3 Dynamics

A schematic plot of a typical cavity-assisted optomechanical device is shown in Fig. 5.1. In the limit of a large cavity bandwidth $\gamma \gg \omega_m$, with the cavity mode adiabatically eliminated, the input-output relation for the optical field becomes,

$$\hat{b}_1(t) = \hat{a}_1(t), \tag{5.7}$$

$$\hat{b}_2(t) = \hat{a}_2(t) + (\alpha/\hbar)\hat{x}(t). \tag{5.8}$$

Here $\hat{a}_{1,2}(\hat{b}_{1,2})$ are the input (output) amplitude and phase quadratures; α determines the interaction strength, and it is related to the optical power I_0 by $\alpha \equiv 8\sqrt{2}(\mathcal{F}/\lambda)\sqrt{\hbar I_0/\omega_0}$ where $\mathcal{F}$ is the cavity finesse, and ω_0 the laser frequency.

The equations of motion for the oscillator are

$$\dot{\hat{x}}(t) = \hat{p}(t)/m, \tag{5.9}$$

$$\dot{\hat{p}}(t) = -m\omega_m^2 \hat{x}(t) + \alpha \hat{a}_1(t) + \hat{\xi}_{\text{th}}(t). \tag{5.10}$$

Here, we can split the oscillator position into a perturbed part $\delta \hat{x}$, and a free oscillation part $\hat{x}_q$, which is given by:

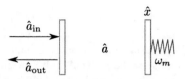

Fig. 5.1 A schematic plot showing an optomechanical system. A mechanical oscillator with eigenfrequency ω_m interacts with a cavity mode $\hat{a}$, which is also coupled to the input ($\hat{a}_{\rm in}$) and output ($\hat{a}_{\rm out}$) optical modes

$$\hat{x}_q(t) = \delta x_q[\hat{X}_0 \cos\omega_m t + \hat{P}_0 \sin\omega_m t], \tag{5.11}$$

where $\delta x_q \equiv \sqrt{\hbar/(2m\omega_m)}$. The perturbed part includes the effects of the radiation-pressure noise, thermal noise, and external driving force (our signal), namely:

$$\delta\hat{x}(t) = \int_0^t dt' G_x(t-t')[\alpha\hat{a}_1(t') + \xi_{\rm th}(t') + F_{\rm ext}(t')], \tag{5.12}$$

where $G_x(t) \equiv \Theta(t)\sin\omega_m t/(m\omega_m)$ denotes the Green's function.

5.4 Variational Quadrature Readout

In order to make the detection scheme work for all possible signals, we need to discretize the measurement process into small time sections, the sum of which tells us the shape of the signal. This idea is similar to the discrete sampling variational measurement discussed in Ref. [4, 5], in which a back-action-evading scheme for a free mass is proposed. Here, we try to derive the noise spectrum for this scheme, and focus on the oscillator case.

Suppose we divide the measurement from $t = 0$ to $t = T$ into N sections (the spectral resolution bandwidth is given by $1/T$). For simplicity, we assume that each section has the same duration $\Delta\tau$, namely $\Delta\tau = T/N$. With time-dependent homodyne detection, we can construct the following N integral estimators for the output signal:

$$\hat{Y}_i = \int_{\tau_i}^{\tau_{i+1}} dt \left[g_1^{(i)}(t)\hat{b}_1(t) + g_2^{(i)}(t)\hat{b}_2(t)\right], \quad (\tau_i = i\Delta\tau,\ i = 0, 1, \cdots, N-1). \tag{5.13}$$

In order to form a quasi-QND measurement, each section should measure the same quadrature. This is because:

$$[\hat{Y}_i, \hat{Y}_j] = 0 (i \neq j), \quad [\hat{X}_\zeta, \hat{X}_{\zeta'}] \neq 0. \tag{5.14}$$

That is, different quadratures are probed, the added noise between different sections will be Heisenberg-limited. This requirement can be trivially satisfied if τ is equal to

5.4 Variational Quadrature Readout

an integer times the oscillation period, and g_2 is the corresponding periodic function. One simple choice would be:

$$\Delta\tau = \tau_m = 2\pi/\omega_m, \quad g_2^{(j)}(t) = g_2^{(i)}(t + j\tau_m - i\tau_m). \tag{5.15}$$

In this case, the filtering function has the following Fourier decomposition:

$$g_2(t) = \frac{1}{\tau_m}\sum_{n=-\infty}^{+\infty} \tilde{g}_{2,n}(n\omega_m) e^{-in\omega_m t}, \quad \tilde{g}_{2,n}(n\omega_m) = \int_0^{\tau_m} dt\, g_2(t) e^{in\omega_m t}. \tag{5.16}$$

We encounter one immediate problem with this scheme: each section only senses $\hat{a}_1$ for one oscillation period, while the effects of back-action accumulate from $t = 0$. The back-action of each section can not be evaded, if the filtering functions only satisfy the back-action-evading (BAE) condition [cf. Eq. (5.18)]. As it turns out, we have to sacrifice the sensitivity around the mechanical resonance. This is because of the following argument: by swapping the integration order, the back-action part of the estimator:

$$\delta\hat{Y}_i = \int_{\tau_i}^{\tau_{i+1}} dt\, g_1^{(i)}(t)\hat{a}_1(t) + (\alpha^2/\hbar) \int_{\tau_i}^{\tau_{i+1}} dt\, \hat{a}_1(t) \int_t^{\tau_{i+1}} dt'\, G_x(t'-t) g_2^{(i)}(t')$$
$$+ (\alpha^2/\hbar) \int_0^{\tau_i} dt\, \hat{a}_1(t) \int_{\tau_i}^{\tau_{i+1}} dt'\, G_x(t'-t) g_2^{(i)}(t'). \tag{5.17}$$

If the BAE condition is satisfied, namely:

$$g_1^{(i)}(t) + (\alpha^2/\hbar) \int_t^{\tau_{i+1}} dt'\, G_x(t'-t) g_2^{(i)}(t') = 0. \tag{5.18}$$

We have

$$\delta\hat{Y}_i = (\alpha^2/\hbar) \int_0^{\tau_i} dt\, \hat{a}_1(t) \int_{\tau_i}^{\tau_{i+1}} dt'\, G_x(t'-t) g_2^{(i)}(t'). \tag{5.19}$$

Therefore, to eliminate this accumulated back-action, we require $\int_{\tau_i}^{\tau_{i+1}} dt'\, G_x(t'-t) g_2^{(i)}(t') = 0$ for any t, which automatically leads to:

$$\int_{\tau_i}^{\tau_{i+1}} dt\, g_2^{(i)}(t)\hat{x}_q(t) = 0. \tag{5.20}$$

If this is the case, we obtain:

$$\hat{Y}_i^{\text{BAE}} = \int_{\tau_i}^{\tau_{i+1}} dt\, g_2^{(i)}(t)[\hat{a}_2(t) + (\alpha/\hbar)\int_{\tau_i}^{t} dt'\, G_x(t-t') F_{\text{ext}}(t')]. \tag{5.21}$$

We can recover the sensitivity around ω_m by using stroboscopic variation measurement, as we will discuss in Sect. 5.5. Even though we do not have resonance gain, this scheme could still be interesting if the signals that we are searching for do not have frequency components at ω_m.

In the following, we will derive the noise spectrum for such a scheme. Using these BAE estimators, we can construct the output spectrum for the entire measurement process; we have:

$$\hat{Z}_n = \tau_m \sum_{k=0}^{N-1} \hat{Y}_k^{\text{BAE}} e^{i n \omega_e \tau_k}, \quad \omega_e \equiv 2\pi/T. \tag{5.22}$$

The corresponding spectrum S_{ZZ} would be defined through

$$\langle \hat{Z}_n \hat{Z}_{n'} \rangle = S_{ZZ} \delta_{nn'} N \tau_m. \tag{5.23}$$

Substituting the expressions for g_2 and $\hat{Y}_k^{\text{BAE}}$ into the above equation, and after some straightforward calculations, we obtain:

$$S_{ZZ}(n\omega_e) = \sum_{k=-\infty}^{\infty} |\tilde{g}_k(k\omega_m)|^2 \left[\frac{1}{2} + \frac{\alpha^2}{\hbar^2} \frac{4\sin^2(n\omega_e \tau_m/2)}{(n\omega_e \tau_m + 2k\pi)^2} R_{xx}^2(n\omega_e) S_F(n\omega_e) \right]$$
$$(\tilde{g}_k(\pm\omega_m) = 0), \tag{5.24}$$

where the mechanical response function $R_{xx}(\omega) = [m(\omega^2 - \omega_m^2)]^{-1}$. Taking the limit $T \to \infty$ (i.e., a very good bandwidth resolution), we have:

$$S_{ZZ}(\Omega) = \sum_{k=-\infty}^{\infty} |\tilde{g}_k(k\omega_m)|^2 \left[\frac{1}{2} + \frac{\alpha^2}{\hbar^2} \frac{\sin^2(\omega \tau_m/2)}{[(\omega \tau_m/2) + k\pi]^2} R_{xx}^2(\Omega) S_F(\Omega) \right], \tag{5.25}$$

where q is the squeezing factor—a 10 dB squeezing gives $e^{-2q} = 0.1$; $S_F(\Omega)$ is the force noise spectrum that contains both thermal noise, and the external driving force. This result is intuitively expected. Basically, the displacement sensitivity is the original one multiplied by the sinc function due to the discrete sampling. Therefore, if the GW signal has a characteristic frequency much lower than the mechanical frequency, the filtering function g_2 should be mostly a constant. In this case, the normalized noise is given by:

$$S_h(\Omega) = \frac{1}{m^2 L^2 \Omega^4} \left[\frac{\hbar^2}{2\alpha^2} \frac{(\Omega \tau_m/2)^2}{\sin^2(\Omega \tau_m/2)} R_{xx}^{-2}(\Omega) + S_F^{\text{th}}(\Omega) \right]. \tag{5.26}$$

Even though this is shot-noise limited, this sensitivity does not really increase a lot at low frequencies, because it rises as $1/\Omega^2$ and physically we know that the oscillator does not has a good response at low frequencies. However, we show that back-action can be evaded simply by manipulating the output, for all possible signals.

5.5 Stroboscopic Variational Measurement

In order to recover the sensitivity around the mechanical resonance, we have to devote most of the measurement time to the amplitude quadrature, and only measure the phase quadrature for a short period of time $\Delta t \ll \omega_m^{-1}$ every oscillator cycle —the idea of stroboscopic variational measurement. This is different from what is proposed in Ref. [4] and here the driving field is still constant in time.

As in previous considerations, we divide the measurement into N sections with $\Delta \tau = \tau_m$. Instead of measuring both amplitude and phase quadrature simultaneously, we measure the phase quadrature during $[\tau_i, \tau_i + \Delta t]$, and switch to the amplitude quadrature during $[\tau_i + \Delta t, \tau_{i+1}]$. Mathematically, we have the following $2N$ integral estimators:

$$\hat{X}_i = \int_{\tau_i + \Delta t}^{\tau_{i+1}} dt \, g_1^{(i)}(t) \hat{b}_1(t), \quad \hat{Y}_i = \int_{\tau_i}^{\tau_i + \Delta t} dt \, g_2^{(i)}(t) \hat{b}_2(t). \quad (5.27)$$

To minimize the back-action, we can construct the following quantities:

$$\hat{O}_i \equiv \sum_{n=0}^{i-1} \hat{X}_n(t) + \hat{Y}_i(t). \quad (5.28)$$

Basically, for each phase quadrature readout, we take into account all the back-action imposed during the history. One can easily find that as long as $g_1(t)$ and $g_2(t)$ are periodic functions at the mechanical frequency and

$$g_1^{(i)}(t) + \int_{\tau_i}^{\tau_i + \Delta t} dt' G_x(t' - t) g_2^{(i)}(t') = 0, \quad (t < \tau_i) \quad (5.29)$$

we have

$$\hat{O}_i = \int_{\tau_i}^{\tau_i + \Delta t} dt \, g_2^{(i)}(t) \hat{a}_2(t) + (\alpha/\hbar) g_2^{(i)}(t) \int_0^t dt' G_x(t - t') [\alpha H(t') \hat{a}_1(t') + F_{\text{ext}}(t')], \quad (5.30)$$

where $H(t)$ is a periodic function of τ_m and $H(t) = \Theta(t) - \Theta(t - \Delta t)$ during $[0, \tau_m]$. Therefore, most of the back actions are evaded, and the remaining part is imposed during the period when we measure the phase quadrature. This is exactly the same as in the stroboscopic measurement, but with an important difference that the probing light beam is always on in our case, which is more feasible experimentally. By applying the same procedure as in the previous case, we can evaluate the output spectrum from:

$$\hat{Z}_n = \tau_m \sum_{k=1}^{N-1} (\hat{O}_k - \hat{O}_0) e^{in\omega_e \tau_k}. \quad (5.31)$$

Here we remove the initial condition term of the oscillator position by subtracting $\hat{O}_k$ with $\hat{O}_0$. After some manipulations, and expanding the results in series of $\omega_m \Delta t$, we find:

$$S_{ZZ}(n\omega_e) = \sum_{k=-\infty}^{\infty} |\tilde{g}_k(k\omega_m)|^2 \frac{\Delta t}{2\tau_m}$$

$$+ \sum_{k=-N_0}^{N_0} |\tilde{g}_k(k\omega_m)|^2 \left[\frac{\alpha^4 \Delta t^5}{24\hbar^2 m^2 \tau_m \sin^2(n\omega_e \tau_m/2)} + \frac{\alpha^2 \Delta t^2}{2\hbar^2 \tau_m^2} R_{xx}^2(n\omega_e) S_F(n\omega_e) \right],$$
(5.32)

where $N_0 \omega_m \Delta t \ll 1$ so that the Taylor expansion is justified, and terms with larger k are small and do not contribute to the summation. Suppose g_2 only has a frequency component at ω_m. Taking the continuous limit, the normalized spectral density for detecting GWs is given by:

$$S_h(\Omega) = \frac{1}{m^2 L^2 \Omega^4} \left[\frac{\hbar^2 \tau_m}{\alpha^2 R_{xx}^2(\Omega) \Delta t} + \frac{\alpha^2 \tau_m \Delta t^3}{12 m^2 R_{xx}^2(\Omega) \sin^2(\Omega \tau_m/2)} + S_F^{\text{th}}(\Omega) \right].$$
(5.33)

Due to the presence of the sin factor in the denominator, the back-action noise never vanishes, even at the mechanical resonance. However, the added noise in principle can be arbitrarily small, if the optimal time scale is chosen, and the measurement is sufficiently sensitive.

5.6 Conclusions

We have considered two types of variational measurement schemes for measuring the QND variable-the mechanical quadrature. In both cases, there are insignificant contributions from the back-action noise to the sensitivity. In contrast to the approach of modulating the interaction strength, and requiring a good-cavity condition, we have considered the case with a large cavity bandwidth. Therefore, these schemes are more suitable for small-scale cavity-assisted devices and large-scale broadband GW detectors.

References

1. V.B. Braginsky, F.Y. Khalili, *Quantum Measurement* (Cambridge University Press, Cambridge, 1992)
2. V.B. Braginsky, Y.I. Vorontsov, K.S. Thorne, Quantum nondemolition measurements. Science **209**, 547–557 (1980)
3. A.A. Clerk, F. Marquardt, K. Jacobs, Back-action evasion and squeezing of a mechanical resonator using a cavity detector. New J. Phys. **10**(9), 095010 (2008)

References

4. S.L. Danilishin, F.Y. Khalili, S.P. Vyatchanin, The discrete sampling variation measurement. Phys. Lett. A **278**, 123–128 (2000)
5. S.L. Danilishin, F.Y. Khalili, Stroboscopic variation measurement. Phys. Lett. A **300**, 547–558 (2002)
6. F.J. Giessibl, Advances in atomic force microscopy. Rev. Mod. Phys. **75**, 949 (2003)
7. S.P. Vyatchanin, A.B. Matsko, Quantum limit on force measurements. JETP **77**, 218 (1993)
8. S.P. Vyatchanin, E.A. Zubova, Quantum variation measurement of a force. Phys. Lett. A **201**, 269–274 (1995)
9. http://www.ligo.caltech.edu/advLIGO

Chapter 6
MQM With Three-Mode Optomechanical Interactions

6.1 Preface

In this chapter, we discuss the Macroscopic Quantum Mechanics (MQM) of a three-mode optomechanical system, in which two orthogonal transverse optical cavity modes are coupled to one mechanical mode through radiation pressure. This work is motivated by the investigations of three-mode parametric instability in large-scale gravitational-wave (GW) detectors with high-power optical cavities, as first pointed out by Braginsky et al. [3]. We realized that the same mechanism that induces instability, in a different parameter regime, can also be used to cool the mechanical resonator down to its quantum ground state. Different from the classical analysis by Braginsky et al. we present a full quantum analysis of three-mode optomechanical parametric interactions, which properly takes into account quantum fluctuations and correlations. We obtain the quantum limit for the ground state cooling with three-mode interactions. In addition, we show that it can also create tripartite optomechanical quantum entanglement between the cavity modes and the mechanical oscillator. Compared with the conventional cavity-assisted optomechanical devices using a single cavity mode, three-mode interactions can achieve an optimal frequency matching: the frequency separation of the two cavity modes is equal to the mechanical-mode frequency. This allows the carrier and sideband fields to simultaneously resonate and coherently build up. Such a mechanism significantly enhances the optomechanical couplings in the quantum regime. It allows us to explore quantum behaviors of optomechanical interactions in small-scale table-top experiments. We show explicitly that given experimentally achievable parameters, three-mode scheme can realize quantum ground-state cooling of milligram scale mechanical oscillators and create robust stationary tripartite optomechanical quantum entanglements. This chapter summarizes a joint research effort by Chunnong Zhao, Li Ju, David Blair, Zhongyang Zhang and me. The relevant publications are Phys. Rev. A **78**, 063809 (2008), Phys. Rev. A **79**, 063801 (2009), andu Phys. Rev. Letts. **104**, 243902 (2009).

6.2 Introduction

Optomechanical interactions have recently become of great interest, for their potential in exploring the quantum behavior of macroscopic objects. Various experiments have demonstrated that the mechanical mode of a mechanical oscillator can be cooled significantly through two-mode optomechanical interactions [2, 7, 8, 9, 12, 17, 24, 27, 30, 33, 34, 38]. The basic setup consists of a Fabry-Pérot cavity with an end mirror. Linear oscillations of the mirror mechanical mode at frequency ω_m scatter the optical-cavity mode (usually TEM_{00}) at frequency ω_0 into Stokes ($\omega_0 - \omega_m$) and anti-Stokes ($\omega_0 + \omega_m$) sideband modes, which have the same spatial mode shape as the TEM_{00} mode. The optical cavity is appropriately detuned such that the anti-Stokes sideband is close to resonance. Therefore, the anti-Stokes process is favored over the Stokes process. As a natural consequence of energy conservation, the thermal energy of the mechanical mode has to decrease in order to create higher-energy anti-Stokes photons at $\omega_0 + \omega_m$. If the cavity-mode decay rate, which is related to the optical finesse, is smaller than the mechanical-mode frequency, theoretical analysis shows that these experiments can eventually achieve the quantum ground state of a macroscopic mechanical oscillator [11, 20, 41], which would be a significant breakthrough in physics from both experimental and theoretical points of view. With the same scheme, many interesting issues have been raised in the literature, such as teleportation of a quantum state into mechanical degrees of freedom [19], creation of stationary quantum entanglements between the cavity mode and the mechanical oscillator [28, 40], or even between two oscillators [15, 26]. This in turn could be implemented in future quantum communications and computing.

The concept of three-mode optomechanical parametric interactions was first introduced and analyzed theoretically in the pioneering work of Braginsky et al. [3]. It was shown that three-mode interactions inside high-power optical cavities of large-scale laser interferometric gravitational-wave (GW) detectors have the potential to induce instabilities, which would severely undermine the operation of detectors. This analysis was elaborated by many other authors to more accurately simulate the real situation in next-generation advanced gravitational-wave detectors [4, 13, 16] and to find strategies for suppressing instability [5, 13]. Recently, the UWA group experimentally demonstrated three-mode interactions in an 80-m high-power optical cavity by exciting the mechanical modes and observing resonant scattering of light into a transverse cavity mode [43].

In contrast to the two-mode case, in three-mode interactions, a single mechanical mode of the mechanical oscillator scatters the main cavity TEM_{00} mode into another transverse cavity mode, which has a different spatial distribution from the TEM_{00} mode. Specifically, when the TEM_{00} mode is scattered by the mechanical mode, the frequency is split into Stokes and anti-Stokes sidebands at $\omega_0 \pm \omega_m$, and in addition, the spatial wavefront is also modulated by the mechanical mode. Three-mode interactions strongly arise when both the modulation frequency and spatial mode distribution are closely matched to those of another transverse optical-cavity mode. Under these circumstances, both the carrier and sideband modes are simultaneously resonant inside the cavity and get coherently built up. Taking into account the resonance of the mechanical mode, the system is triply resonant, with the interaction strength scaled by the product of the two optical quality factors and the mechanical quality factor. If the transverse optical cavity mode has a frequency lower than the TEM_{00} mode, the Stokes sideband will be on resonance and the interaction provides positive amplification of the mechanical mode, while if the transverse mode has a frequency above the main cavity TEM_{00} mode, the anti-Stokes sideband will resonate, and the system has negative gain and the mechanical mode will be cooled. The underlying principle of both two and three mode interactions is similar to the Brillouin scattering, except that the modulation occurs not through changes in refractive index of the medium, but through bulk surface motion of a macroscopic mechanical oscillator (i.e. the mechanical mode) which modulates the optical path of the light.

While three-mode interactions are inconvenient by-products of the design of advanced GW detectors, they can be engineered to occur in small-scale systems with low mass resonators, which can serve as an optomechanical amplifier and be applied to mechanical-mode cooling [23, 42]. Besides, due to its triply resonant feature, the three-mode system has significant advantages compared with the two-mode system and allows much stronger optomechanical couplings. To motivate experimental realizations, we have suggested a small-scale table-top experiment with a milligram mechanical oscillator in a coupled cavity [42]. Using the extra degree of freedom of the coupled cavity, the cavity mode gap (i.e., the difference between the two relevant cavity modes) can be continuously tuned such that it is equal to plus or minus the mechanical-mode frequency, which maximizes the three-mode interaction strength. We also pointed out that, in the negative-gain regime, this experimental setup can be applied to resolved-sideband cooling of a mechanical oscillator down to its quantum ground state. In that paper, we used the classical analysis presented by Braginsky et al. to obtain the effective thermal occupation number $\bar{n}$ of the mechanical mode. This analysis *breaks down* when $\bar{n} \ll 1$ and the quantum fluctuations of the cavity modes have to be taken into account. To overcome this limitation, we used the similarity in the Hamiltonian of the two-mode and the three-mode system, and argued that the quantum limit for cooling in both systems is the same without investigating the detailed dynamics. However, in order to gain a quantitative understanding of the three-mode system in the quantum regime, it is essential to develop a full quantum analysis which includes the dynamical effects of the quantum fluctuations. Besides, as we will show, the quantum analysis reveals a most interesting non-classical feature of three-mode systems: stationary tripartite quantum entanglement.

The outline of this chapter is as follows: in Sect. 6.3, we start from the classical analysis given by Braginsky et al. and then quantize it with the standard approach. In Sect. 6.4, we use the quantized Hamiltonian as the starting point to analyze the dynamics of the three-mode system. Further, based upon the Fluctuation-Dissipation-Theorem (FDT), we derive the quantum limit for the achievable thermal occupation number in cooling experiments. To motivate future small-scale experiments, we provide an experimentally achievable specification for the quantum ground state cooling of a mechanical oscillator. In Sect. 6.5, we investigate the stationary tripartite

Fig. 6.1 Spatial shapes of the TEM_{00}, and of TEM_{01} modes and the mechanical torsional mode. From Ref. [3]

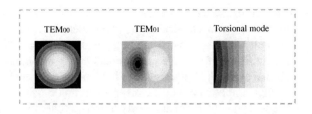

optomechanical quantum entanglement, and we show that the same specification for the cooling experiments can also be applied to realize robust stationary optomechanical entanglements.

6.3 Quantization of Three-Mode Parametric Interactions

In this section, we will first present the classical formulations of three-mode optomechanical parametric interactions given by Braginsky et al. [3], and then apply standard procedures to obtain the quantized version.

Classical Picture. A detailed quantitative classical formulation of three-mode interactions was given in the Appendix of Ref. [3]. A Lagrangian formalism was used to derive the classical equations of motion and analyze the stability of the entire three-mode optomechanical system. The formalism can be easily converted into Hamiltonian language, which can then be quantized straightforwardly. For convenience, we will use slightly different notation and definitions for the optical fields. Further, we assume that the two optical-cavity modes are the TEM_{00} and TEM_{01} modes, and that the mechanical mode has a torsional mode shape (about the vertical axis) which has a large spatial overlap with the TEM_{01} mode as shown in Fig. 6.1. This can be easily extended to general cases with other transverse optical modes and mechanical modes. Assuming the electric field is linearly polarized in the transverse direction perpendicular to the z axis, the electromagnetic fields (E, H) of the cavity modes can be written as:

$$E_i(t) = \left(\frac{\hbar \omega_i}{\epsilon_0 V}\right)^{1/2} f_i(\vec{r}_\perp) \sin(k_i z) q_i(t), \tag{6.1}$$

$$H_i(t) = \frac{\epsilon_0}{k_i} \left(\frac{\hbar \omega_i}{\epsilon_0 V}\right)^{1/2} f_i(\vec{r}_\perp) \cos(k_i z) \dot{q}_i(t). \tag{6.2}$$

Here, $i = 0, 1$ represent the TEM_{00} and TEM_{01} modes; $f_i(\vec{r}_\perp)$ are the transverse mode shapes; ω_i denotes the eigenfrequency; k_i are the wave numbers; V is the volume of the optical cavity; $q_i(t)$ are the generalized coordinates of the fields; and $\dot{q}_i(t)$ are the time derivatives of $q_i(t)$. At the present stage, the appearance of $\hbar \omega_i$ is just to make the generalized coordinates $\hat{q}_i$ dimensionless. The classical Hamiltonian of this system is given by:

$$\mathcal{H} = \mathcal{H}_m + \frac{1}{2} \int d\vec{r}_\perp (L + x u_z) [\epsilon_0 (E_0 + E_1)^2 + \mu_0 (H_0 + H_1)^2], \tag{6.3}$$

6.3 Quantization of Three-Mode Parametric Interactions

where L is the length of the cavity; x is the generalized coordinate of the mechanical mode; u_z is the vertical displacement. The free Hamiltonian of the mechanical mode is

$$\mathcal{H}_m = \frac{1}{2}\hbar\omega_m(q_m^2 + p_m^2) \tag{6.4}$$

with $q_m \equiv x/\sqrt{\hbar/(m\omega_m)}$, and p_m is the momentum normalized with respect to $\sqrt{\hbar m \omega_m}$.

After integrating over the transverse direction, and taking into account the mode shapes, we obtain:

$$\mathcal{H} = \mathcal{H}_m + \mathcal{H}_0 + \mathcal{H}_1 + \mathcal{H}_{\text{int}}. \tag{6.5}$$

Defining dimensionless canonical momentum $p_i(t) \equiv \dot{q}_i(t)/\omega_i$, the free Hamiltonian of the two cavity modes are:

$$\mathcal{H}_i = \frac{1}{2}\hbar\omega_i(q_i^2 + p_i^2) \tag{6.6}$$

and the interaction Hamiltonian is given by

$$\mathcal{H}_{\text{int}} = \hbar G_0 q_m (q_0 q_1 + p_0 p_1), \tag{6.7}$$

where the coupling constant is defined as $G_0 \equiv \sqrt{\Lambda \hbar \omega_0 \omega_1/(m \omega_m L^2)}$, with the geometrical overlapping factor $\Lambda \equiv (L \int d\vec{r}_\perp u_z f_0 f_1/V)^2$.

Given the above Hamiltonian, it is straightforward to derive the classical equations of motion and analyze the dynamics of the system, which would be identical to those in the Appendix of Ref. [3]. To quantify the strength of three-mode interactions, Braginsky et al. introduced the parametric gain $\mathcal{R}$, as defined by:

$$\mathcal{R} = \pm \frac{2\Lambda I_0 \omega_1}{m \omega_m L^2 \gamma_0 \gamma_1 \gamma_m} = \pm \frac{2\Lambda I_0 Q_0 Q_1 Q_m}{m \omega_0 \omega_m^2 L^2}, \tag{6.8}$$

where $\pm$ correspond to either positive gain or negative gain; I_0 is the input optical power of the TEM$_{00}$ mode; and we have defined optical and mechanical-mode quality factors $Q_i = \omega_i/\gamma_i$ ($i = 0, 1, m$). Due to optomechanical interaction, the decay rate γ_m of the mechanical mode will be modified to an effective one γ_m', which is

$$\gamma_m' \approx (1 - \mathcal{R})\gamma_m. \tag{6.9}$$

When $\mathcal{R} > 1$, the decay rate becomes negative and this corresponds to instability. Here we are particularly interested in the regime where $\mathcal{R} < 0$ which gives rise to the mechanical-mode cooling. The effective thermal occupation number $\bar{n}_{\text{th}}'$ of the mechanical mode is given by

$$\bar{n}_{\text{th}}' = \frac{\bar{n}_{\text{th}} \gamma_m}{\gamma_m'} = \frac{\bar{n}_{\text{th}}}{1 - \mathcal{R}} \tag{6.10}$$

with $\bar{n}_{\text{th}}$ denoting the original thermal occupation number. It seems that when $-\mathcal{R} \gg 1$, $\bar{n}'_{\text{th}}$ can be arbitrarily small. However, in this case, the classical analysis breaks down and the quantum fluctuations of the cavity modes will set forth a quantum limit for the minimally achievable $\bar{n}'_{\text{th}}$, which will be detailed in the following quantum analysis.

Quantized Hamiltonian. The classical Hamiltonian derived above can be quantized by identifying the generalized coordinate and momentum as Heisenberg operators, which satisfy the following commutation relations:

$$[\hat{q}_j, \hat{p}_{j'}] = i\,\delta_{jj'}, \quad (j, j' = 0, 1, m). \tag{6.11}$$

The quantized Hamiltonian is then given by

$$\hat{\mathcal{H}} = \frac{1}{2}\sum_{i=m,0,1} \hbar\omega_i(\hat{q}_i^2 + \hat{p}_i^2) + \hbar G_0 \hat{q}_m(\hat{q}_0\hat{q}_1 + \hat{p}_0\hat{p}_1) + \hat{\mathcal{H}}_{\text{ext}}, \tag{6.12}$$

where we have added $\hat{\mathcal{H}}_{\text{ext}}$ to take into account the coupling between cavity modes and external continuum optical fields due to the non-zero transmission of the cavity. This Hamiltonian is convenient for discussing stationary tripartite quantum entanglement as will be shown in Sect. 6.5, as these generalized coordinates q_i and p_i correspond to the amplitude and phase quadratures in the quantum optics entanglement experiments.

To discuss the ground-state cooling as will be investigated in Sect. 6.4, it is illuminating to introduce annihilation operators for the two cavity modes $\hat{a} \equiv (\hat{q}_0 + i\,\hat{p}_0)/\sqrt{2}$ and $\hat{b} \equiv (\hat{q}_1 + i\,\hat{p}_1)/\sqrt{2}$, so that the normally-ordered quantized Hamiltonian can be rewritten as:

$$\hat{\mathcal{H}} = \frac{1}{2}\hbar\omega_m(\hat{q}_m^2 + \hat{p}_m^2) + \hbar\omega_0\hat{a}^\dagger\hat{a} + \hbar\omega_1\hat{b}^\dagger\hat{b} + \hbar G_0\hat{q}_m(\hat{a}^\dagger\hat{b} + \hat{b}^\dagger\hat{a}) + \hat{\mathcal{H}}_{\text{ext}}. \tag{6.13}$$

6.4 Quantum Limit for Three-Mode Cooling

In this section, we will start from the Hamiltonian in Eq. (6.13) to derive the dynamics and discuss the quantum limit for the ground-state cooling experiments using three-mode optomechanical interactions. As we will see, due to similar mathematical structure as in the two-mode case, the corresponding quantum limit for three-mode cooling is identical to the resolved-sideband limit derived by Marquardt et al. [15] and Wilson-Rae et al. [39] in the two-mode case.

Equations of Motion. The dynamics of this three-mode system can be derived from the quantum Langevin equations (QLEs). In the experiments, the TEM$_{00}$ mode is driven on resonance at ω_0. Therefore, we choose a rotating frame at ω_0, obtaining the corresponding nonlinear QLEs as:

6.4 Quantum Limit for Three-Mode Cooling

$$\dot{\hat{q}}_m = \omega_m \hat{p}_m, \tag{6.14}$$

$$\dot{\hat{p}}_m = -\omega_m \hat{q}_m - \gamma_m \hat{p}_m - G_0(\hat{a}^\dagger \hat{b} + \hat{b}^\dagger \hat{a}) + \xi_{\text{th}}, \tag{6.15}$$

$$\dot{\hat{a}} = -\gamma_0 \hat{a} - i G_0 \hat{q}_m \hat{b} + \sqrt{2\gamma_0} \hat{a}_{\text{in}}, \tag{6.16}$$

$$\dot{\hat{b}} = -(\gamma_1 + i \Delta)\hat{b} - i G_0 \hat{q}_m \hat{a} + \sqrt{2\gamma_1} \hat{b}_{\text{in}}. \tag{6.17}$$

Here the TEM$_{00}$ and TEM$_{01}$ mode gap is given by $\Delta \equiv \omega_1 - \omega_0$; $G_0(\hat{a}^\dagger \hat{b} + \hat{b}^\dagger \hat{a})$ corresponds to the radiation pressure which modifies the dynamics of the mechanical mode and is also responsible for the quantum limit; we have added thermal noise ξ_{th} whose correlation function, in the Markovian approximation, is given by $\langle \xi_{\text{th}}(t)\xi_{\text{th}}(t')\rangle = 2\gamma_m \bar{n}_{\text{th}}\delta(t - t')$. In obtaining the above equations, we have also used the rotating-wave approximation for $\hat{\mathcal{H}}_{\text{ext}}$, namely:

$$\hat{\mathcal{H}}_{\text{ext}} = i\hbar(\sqrt{2\gamma_0}\,\hat{a}^\dagger \hat{a}_{\text{in}} + \sqrt{2\gamma_1}\,\hat{b}^\dagger \hat{b}_{\text{in}} - \text{H.c.}) \tag{6.18}$$

with H.c. denoting the Hermitian conjugate.

To solve the above equations, we can linearize them by replacing every Heisenberg operator with the sum of a steady part and a small perturbed part, namely $\hat{o} = \bar{o} + \delta\hat{o}(\epsilon)$ with $\epsilon \ll 1$. We treat the Brownian thermal noise ξ_{th}, the vacuum fluctuations $\sqrt{\gamma_0}\delta\tilde{a}_{\text{in}}$, $\sqrt{\gamma_1}\delta\tilde{b}_{\text{in}}$ and $\delta\hat{q}_m$ as being of the order of ϵ. In the experiments, the TEM$_{00}$ mode is pumped externally with a large classical amplitude $\bar{a}_{\text{in}}$ while the TEM$_{01}$ mode is not with $\bar{b}_{\text{in}} = 0$. Therefore, to the zeroth order of ϵ, the steady part of the cavity modes are simply given by

$$\bar{a} = \sqrt{2/\gamma_0}\,\bar{a}_{\text{in}} = \sqrt{2I_0/(\gamma_0 \hbar \omega_0)}, \quad \bar{b} = -iG_0 \bar{a} \bar{q}_m. \tag{6.19}$$

Without loss of generality, we can set $\bar{q}_m = 0$. Therefore, $\bar{b} = 0$ and this allows us to eliminate the TEM$_{00}$ mode from the first-order equations, which are:

$$\delta\dot{\hat{q}}_m = \omega_m\,\delta\hat{p}_m, \tag{6.20}$$

$$\delta\dot{\hat{p}}_m = -\omega_m\,\delta\hat{q}_m - \gamma_m\,\delta\hat{p}_m - G_0 \bar{a}(\delta\hat{b} + \delta\hat{b}^\dagger) + \xi_{\text{th}}, \tag{6.21}$$

$$\delta\dot{\hat{b}} = -(\gamma_1 + i\Delta)\delta\hat{b} - iG_0 \bar{a}\,\delta\hat{q}_m + \sqrt{2\gamma_1}\,\delta\hat{b}_{\text{in}}. \tag{6.22}$$

Here we have chosen an appropriate phase reference such that $\bar{a}_{\text{in}}$ is real and positive. The above equations can be solved in the frequency domain, namely:

$$\tilde{q}_m(\Omega) = -\frac{\omega_m[\tilde{F}_{\text{rp}}(\Omega) + \tilde{\xi}_{\text{th}}(\Omega)]}{(\Omega^2 - \omega_m^2) + i\gamma_m\Omega}, \tag{6.23}$$

$$\delta\tilde{b}(\Omega) = \frac{G_0\,\bar{a}\,\delta\tilde{q}_m(\Omega) + i\sqrt{2\gamma_1}\delta\tilde{b}_{\text{in}}(\Omega)}{(\Omega - \Delta) + i\gamma_1}, \tag{6.24}$$

where the radiation pressure

$$\tilde{F}_{\rm rp}(\Omega) = \frac{2G_0^2 \bar{a}^2 \Delta \, \delta\tilde{q}_m(\Omega) - 2G_0\bar{a}\sqrt{\gamma_1}[(\gamma_1 - i\Omega)\delta\tilde{q}_2(\Omega) - \Delta\,\delta\tilde{p}_2(\Omega)]}{[(\Omega - \Delta) + i\,\gamma_1][(\Omega + \Delta) + i\,\gamma_1]} \quad (6.25)$$

with the amplitude and phase quadratures $\delta\tilde{q}_2(\Omega) = [\delta\tilde{b}(\Omega) + \delta\tilde{b}^\dagger(-\Omega)]/\sqrt{2}$ and $\delta\tilde{p}_2(\Omega) = [\delta\tilde{b}(\Omega) - \delta\tilde{b}^\dagger(-\Omega)]/(\sqrt{2}i)$. In the expression for $\tilde{F}_{\rm rp}$, the part proportional to $\delta\tilde{q}_m$ is called the optical spring effect. For a high quality-factor oscillator with $\omega_m \gg \gamma_m$, the decay rate γ_m and the eigenfrequency ω_m of the mechanical mode will be modified to new effective values γ_m' and ω_m', as given by:

$$\gamma_m' = \gamma_m + \frac{4G_0^2 \bar{a}^2 \Delta\, \omega_m \gamma_1}{[(\omega_m - \Delta)^2 + \gamma_1^2][(\omega_m + \Delta)^2 + \gamma_1^2]}, \quad (6.26)$$

$$\omega_m' = \omega_m + \frac{G_0^2 \bar{a}^2 \Delta(\omega_m^2 - \Delta^2 - \gamma_1^2)}{[(\omega_m - \Delta)^2 + \gamma_1^2][(\omega_m + \Delta)^2 + \gamma_1^2]}. \quad (6.27)$$

In our case, the TEM$_{00}$ and TEM$_{01}$ mode gap is $\Delta = \omega_1 - \omega_0 = \omega_m$. In the resolved-sideband case with $\gamma_1 \ll \omega_m$, we obtain:

$$\gamma_m' \approx \gamma_m + \frac{G_0^2 \bar{a}^2}{\gamma_1}; \quad \omega_m' \approx \omega_m - \frac{G_0^2 \bar{a}^2}{4\omega_m}. \quad (6.28)$$

If we define the parametric gain as $\mathcal{R} = (\gamma_m - \gamma_m')/\gamma_m$, then in this case

$$\mathcal{R} = -\frac{G_0^2 \bar{a}^2}{\gamma_1 \gamma_m} = -\frac{2\Delta I_0 \omega_1}{m\omega_m L^2 \gamma_0 \gamma_1 \gamma_m}. \quad (6.29)$$

This is identical to Eq. (6.8) in the negative-gain regime, which was obtained from classical analysis by Braginsky et al. [3]. However, in contrast to Eq. (6.10), the resulting thermal occupation number of the mechanical mode is given by

$$\bar{n}_{\rm th}' = \frac{\bar{n}_{\rm th}\gamma_m}{\gamma_m'} + \bar{n}_{\rm quant} = \frac{\bar{n}_{\rm th}}{1 - \mathcal{R}} + \bar{n}_{\rm quant} \quad (6.30)$$

where the extra term $\bar{n}_{\rm quant}$ originates from the vacuum fluctuations in $F_{\rm rp}$, i.e. terms proportional to $\delta\tilde{p}_2$ and $\delta\tilde{q}_2$. Since, in the case of large $\mathcal{R}$, or equivalently strong optomechanical coupling, $\bar{n}_{\rm th}' \approx \bar{n}_{\rm quant}$ and the mechanical mode will finally reach a thermal equilibrium with the cavity modes. The lowest achievable thermal occupation number $\bar{n}_{\rm quant}$ will be determined by this optical heat bath (i.e., cavity mode + external continuum mode).

To derive this quantum limit $\bar{n}_{\rm quant}$, we will apply the Fluctuation-Dissipation-Theorem (FDT). Specifically, given any two quantities $\hat{A}(t)$ and $\hat{B}(t)$ which linearly depend on field strength, we can define the forward correlation function:

$$C_{\hat{A}\hat{B}}(t - t') \equiv \langle \hat{A}(t)\hat{B}(t')\rangle \quad (t > t'), \quad (6.31)$$

6.4 Quantum Limit for Three-Mode Cooling

where $\langle \rangle$ denotes the ensemble average. According to the FDT, we have

$$\frac{S_{\hat{A}\hat{B}}(\Omega) + S_{\hat{A}\hat{B}}(-\Omega)}{S_{\hat{A}\hat{B}}(\Omega) - S_{\hat{A}\hat{B}}(-\Omega)} = \frac{e^{\beta\hbar\Omega} + 1}{e^{\beta\hbar\Omega} - 1} = 2\bar{n}_{\text{eff}}(\Omega) + 1, \qquad (6.32)$$

or equivalently,

$$\bar{n}_{\text{eff}}(\Omega) = \frac{S_{\hat{A}\hat{B}}(-\Omega)}{S_{\hat{A}\hat{B}}(\Omega) - S_{\hat{A}\hat{B}}(-\Omega)}. \qquad (6.33)$$

where $S_{\hat{A}\hat{B}}(\Omega)$ is the power spectral density (Fourier transform of $C_{\hat{A}\hat{B}}$), $\beta = 1/(k_B T_{\text{eff}})$ and the effective thermal occupation number $\bar{n}_{\text{eff}} \equiv 1/(e^{\beta\hbar\Omega} - 1)$. In our case, we can simply substitute A, B with the amplitude of the TEM$_{01}$ mode $\delta\hat{b}$ by fixing $\hat{q}_m = 0$. From Eq. (6.24) and using the fact that for vacuum fluctuation $\langle \delta\tilde{b}_{\text{in}}(\Omega)\delta\tilde{b}_{\text{in}}^{\dagger}(\Omega') \rangle = 2\pi\delta(\Omega - \Omega')$, we obtain:

$$S_{\delta\hat{b}\delta\hat{b}}(\Omega) = \frac{2\gamma_1}{(\Omega - \Delta)^2 + \gamma_1^2}. \qquad (6.34)$$

Since the mechanical mode have a very high intrinsic quality factor ($\omega_m \gg \gamma_m$), the energy transfer between the cavity modes and the mechanical mode only happens around ω_m. Therefore, from Eqs. (6.33) and (6.34), the final quantum limit is given by:

$$\bar{n}_{\text{quant}} \approx \bar{n}_{\text{eff}}(\omega_m) = \left(\frac{\gamma_1}{2\omega_m}\right)^2, \qquad (6.35)$$

where we have used the fact that for the resonant case, $\Delta = \omega_1 - \omega_0 = \omega_m$. To achieve the quantum ground state, i.e. $\bar{n}_{\text{quant}} \sim 0$, we require $\omega_m \gg \gamma_1$ and this is simply the resolved-sideband limit obtained in the pioneering works of Marquardt et al. [20] and Wilson-Rae et al. [41].

The reason why the quantum limit for three-mode cooling is identical to the two-mode case can be readily understood from the fact that the TEM$_{00}$ mode is eliminated from the optomechanical dynamics as shown explicitly in Eqs. (6.20)–(6.22) and we essentially obtain an effective two-mode system. As suggested by Yanbei Chen [private communication], this equivalence can be made more obvious by mapping this three-mode system into a power and signal-recycled laser interferometer, as shown in Fig. 6.2. The TEM$_{00}$ and TEM$_{01}$ modes can be viewed as the common and differential modes in the interferometer, respectively. The torsional mode corresponds to the differential motion of the end mirrors and Δ is equivalent to the detuning of the signal-recycling cavity. In the power- and signal-recycled interferometer, even though there is no high-order transverse optical mode involved, the two degrees of freedom of the power-recycling mirror and the signal recycling mirror enable simultaneous resonances of the carrier and sideband modes, which is achieved naturally with the three-mode optomechanical scheme.

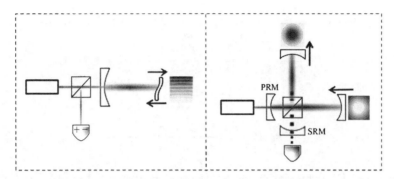

Fig. 6.2 Equivalent mapping from a three-mode system to a power- and signal-recycled interferometer. The TEM$_{00}$ and TEM$_{01}$ modes can be viewed as the common and differential optical modes in the interferometer respectively. The torsional mechanical mode is equivalent to the differential motion of two end mirrors in the interferometer. By adjusting the positions of the power-recycling mirror (*PRM*) and signal-recycling mirror (*SRM*), we can make the carrier and sideband modes simultaneously resonate inside the cavity, the same as in the three-mode scheme. From Ref. [22]

The above discussion shows that, mathematically, two-mode interactions and three-mode interactions are very similar. However, it is very important to emphasize that, from an experimental point of view, there is an important difference. Specifically, the steady-state amplitude $\bar{a}$ in the radiation pressure $F_{\rm rp}$ is amplified by the optical resonance, while for the two-mode case, this amplitude is highly suppressed due to large detuning. In other words, in order to achieve the same optomechanical coupling strength experimentally, the input optical power in the two-mode scheme needs to be $1 + (\Delta/\gamma_0)^2$ times larger than the the three-mode scheme. This is a large factor in the resolved-sideband regime with $\Delta \gg \gamma_0$ (the optimal $\Delta = \omega_m$). Besides, in the three-mode interactions, the condition $\Delta = \omega_m$ also naturally optimizes the energy transfer from the mechanical mode to the cavity mode [10, 20, 41]. Therefore, the three-mode scheme greatly enhances the optomechanical coupling and is able to achieve resolved-sideband limit without compromising the intra-cavity optical power. As mentioned in Ref. [44], the amplitude and laser phase noise can also be reduced significantly with three-mode scheme, simply due to the filtering of the cavity resonance.

To motivate future cooling experiments with three-mode interactions, we now present an experimentally achievable specification for the quantum ground state cooling of a milligram-scale mechanical oscillator. We choose that the mass of the mechanical oscillator $m = 0.1$ mg; the length of the cavity $L = 2$ cm; the mechanical-mode frequency $\omega_m/2\pi = 10^6$ Hz; the mechanical-mode quality factor $Q_m \equiv \omega_m/\gamma_m = 10^7$; and the optical finesse $\mathcal{F} = 10^4$. Given an input optical power of the TEM$_{00}$ mode $I_0 = 50$ mW, and the environmental temperature $T = 4$ K, the corresponding effective thermal occupation number of the mechanical oscillator ~ 0.5.

6.5 Stationary Tripartite Optomechanical Quantum Entanglement

As shown in the works of Vitali et al. [40] and Paternostro et al. [28], optomechanical interaction provides a very efficient way of generating stationary quantum entanglements among cavity modes and the mechanical mode. Once experimentally realized, it will have significant impact on future quantum communications. Following their formalism, we will investigate the stationary tripartite quantum entanglement in the three-mode optomechanical system by first analyzing the dynamics and then evaluating the entanglement measure —the logarithmic negativity $E_\mathcal{N}$ defined in Refs. [1, 39].

Starting from the Hamiltonian in Eq. (6.12), the corresponding nonlinear QLEs in the rotating frame at the laser frequency ω_L can be written as:

$$\dot{\hat{q}}_m = \omega_m \hat{p}_m, \tag{6.36}$$

$$\dot{\hat{p}}_m = -\omega_m \hat{q}_m - \gamma_m \hat{p}_m - G_0(\hat{q}_0 \hat{q}_1 + \hat{p}_0 \hat{p}_1) + \xi_{\text{th}}, \tag{6.37}$$

$$\dot{\hat{q}}_0 = -\gamma_0 \hat{q}_0 + \Delta_0 \hat{p}_0 + G_0 \hat{q}_m \hat{p}_1 + \sqrt{2\gamma_0} \hat{q}_0^{\text{in}}, \tag{6.38}$$

$$\dot{\hat{p}}_0 = -\gamma_0 \hat{p}_0 - \Delta_0 \hat{q}_0 - G_0 \hat{q}_m \hat{q}_1 + \sqrt{2\gamma_0} \hat{p}_0^{\text{in}}, \tag{6.39}$$

$$\dot{\hat{q}}_1 = -\gamma_1 \hat{q}_1 + \Delta_1 \hat{p}_1 + G_0 \hat{q}_m \hat{p}_0 + \sqrt{2\gamma_1} \hat{q}_1^{\text{in}}, \tag{6.40}$$

$$\dot{\hat{p}}_1 = -\gamma_1 \hat{p}_1 - \Delta_1 \hat{q}_1 - G_0 \hat{q}_m \hat{q}_0 + \sqrt{2\gamma_1} \hat{p}_1^{\text{in}}, \tag{6.41}$$

where $\Delta_0 = \omega_0 - \omega_L$, and $\Delta_1 = \omega_1 - \omega_L$. Slightly different from the cooling experiments, here we need to externally drive both the TEM$_{00}$ and TEM$_{01}$ modes simultaneously, to create tripartite quantum entanglement. We choose an appropriate phase reference such that the classical amplitude $\bar{p}_i = 0$ and $\bar{q}_i \neq 0$ ($i = 0, 1$), which is related to the input optical power I_i by $\bar{q}_i = \sqrt{2 I_i /(\hbar \omega_i \gamma_i)}$. Similar to the previous case, we can linearize the above equations as:

$$\dot{\hat{\mathbf{x}}}^{\mathbf{T}} = \mathbf{M} \hat{\mathbf{x}}^{\mathbf{T}} + \hat{\mathbf{n}}^{\mathbf{T}}, \tag{6.42}$$

with **T** denoting the transpose:

$$\hat{\mathbf{x}}^{\mathbf{T}} \equiv (\delta \hat{q}_m, \delta \hat{p}_m, \delta \hat{q}_0, \delta \hat{p}_0, \delta \hat{q}_1, \delta \hat{p}_1)^{\mathbf{T}}, \tag{6.43}$$

$$\hat{\mathbf{n}}^{\mathbf{T}} \equiv (0, \xi_{\text{th}}, \sqrt{2\gamma_0} \delta \hat{q}_0^{\text{in}}, \sqrt{2\gamma_0} \delta \hat{p}_0^{\text{in}}, \sqrt{2\gamma_1} \delta \hat{q}_1^{\text{in}}, \sqrt{2\gamma_1} \delta \hat{p}_1^{\text{in}})^{\mathbf{T}}, \tag{6.44}$$

and matrix **M** is given by:

$$\mathbf{M} = \begin{pmatrix} 0 & \omega_m & 0 & 0 & 0 & 0 \\ -\omega_m & -\gamma_m & G_0 \bar{q}_1 & 0 & G_0 \bar{q}_0 & 0 \\ 0 & 0 & -\gamma_0 & \Delta_0 & 0 & 0 \\ G_0 \bar{q}_1 & 0 & -\Delta_0 & -\gamma_0 & 0 & 0 \\ 0 & 0 & 0 & 0 & -\gamma_1 & \Delta_1 \\ G_0 \bar{q}_0 & 0 & 0 & 0 & -\Delta_1 & -\gamma_1 \end{pmatrix}. \tag{6.45}$$

At first sight, this mathematical structure is identical to the one analyzed by Paternostro et al. [28]. Apart from differing in the coupling constants (here we need to consider the overlapping factor Λ), there is another important difference: after linearization, the radiation pressure term $G_0(\hat{q}_0 \hat{q}_1 + \hat{p}_0 \hat{p}_1)$ in Eq. (6.37) is proportional to $\bar{q}_0 \delta \hat{q}_1 + \bar{q}_1 \delta \hat{q}_0$, rather than to $\bar{q}_0 \delta \hat{q}_0 - \bar{q}_1 \delta \hat{q}_1$, as considered in Ref. [28]. As we will show, similar to the case for cooling experiments, the coherent build-up of both the TEM$_{00}$ and TEM$_{01}$ modes, and the optimal mode gap $\omega_1 - \omega_0 = \omega_m$ enhance the entanglement significantly, which make it easier to achieve experimentally.

Assuming the system is stable, i.e., all eigenvalues of $\mathbf{M}$ have negative real parts, the stationary solutions to Eq. (6.42) can be written formally as:

$$\hat{x}_i(\infty) = \sum_j \int_0^\infty dt' [e^{\mathbf{M}(t-t')}]_{ij} \hat{n}_j(t'), \qquad (6.46)$$

where we have neglected the initial-condition terms, which decay away as the system approaches the stationary state. We assume that all the noises are Markovian Gaussian processes, and that the correlation functions are:

$$\sigma_{ij}(t-t') \equiv D_{ij} \delta(t-t'), \qquad (6.47)$$

where D_{ij} are the elements of matrix $\mathbf{D}$, and

$$D_{ij} = \mathrm{Diag}[0, 2\gamma_m k_B T/(\hbar \omega_m), \gamma_0, \gamma_0, \gamma_1, \gamma_1]. \qquad (6.48)$$

The corresponding stationary covariance matrix among the cavity modes and the mechanical mode can then be written as:

$$\mathbf{V}(\infty) = \int_0^\infty dt [e^{\mathbf{M}t}] \mathbf{D} [e^{\mathbf{M}t}]^\mathbf{T}, \qquad (6.49)$$

and the components of $\mathbf{V}$ can be obtained by solving the following linear equations:

$$\mathbf{MV} + \mathbf{VM^T} = -\mathbf{D}. \qquad (6.50)$$

For this tripartite continuous-variable system (one mechanical mode + two cavity modes), one necessary and sufficient condition for separability is the positivity of partially-transposed covariance matrix [29, 36, 37]. In our case, a partial transpose is equivalent to time reversal and can be realized by reversing the momentum of the mechanical mode from $\hat{p}_m$ to $-\hat{p}_m$, namely

$$\mathbf{V}_{\mathrm{pt}} = \mathbf{V}|_{\hat{p}_m \to -\hat{p}_m}. \qquad (6.51)$$

By evaluating the positivity of the eigenvalue of $\mathbf{V}_{\mathrm{pt}}$, we can directly determine whether entanglement exists or not. To reveal the richness of the entanglement structure, we will not directly analyze the positivity of $\mathbf{V}_{\mathrm{pt}}$ for the entire system, but rather, following Ref. [28], we look at the entanglement between any bipartite subsystem

6.5 Stationary Tripartite Optomechanical Quantum Entanglement

using the logarithmic negativity $E_\mathcal{N}$. Given the 4×4 covariance matrix $\mathbf{V}_{\text{sub}}$ for any bipartite subsystem,

$$\mathbf{V}_{\text{sub}} = \begin{bmatrix} \mathbf{A}_{2\times 2} & \mathbf{C}_{2\times 2} \\ \mathbf{C}^{\mathsf{T}}_{2\times 2} & \mathbf{B}_{2\times 2} \end{bmatrix}, \tag{6.52}$$

the logarithmic negativity $E_\mathcal{N}$ is defined by [1, 29]

$$E_\mathcal{N} = \max[0, -\ln 2\sigma_-] \tag{6.53}$$

with $\sigma_- \equiv \sqrt{\Sigma - \sqrt{\Sigma^2 - 4\det \mathbf{V}_{\text{sub}}}}/\sqrt{2}$, and $\Sigma \equiv \det \mathbf{A} + \det \mathbf{B} - 2\det \mathbf{C}$.

For numerical estimations, we will use the same specification as given in the previous section for the cooling experiments. We will focus on the situation relevant to the experiments, with $\omega_1 - \omega_0 = \omega_m$, and the TEM$_{00}$ mode driven on resonance ($\Delta_0 = 0$, $\Delta_1 = \omega_m$). In Fig. 6.3, we show the resulting $E_\mathcal{N}$ as a function of the input optical powers of both optical modes. Given the specifications, the entanglement strength between each optical mode and the mechanical mode becomes stronger as the optical power of their counterpart increases (until the system becomes unstable). This is understandable, because we have a $\hat{q}_m(\hat{q}_0\hat{q}_1 + \hat{q}_0\hat{q}_1)$ type of interaction, and the coupling strength between the TEM$_{00}$ mode and the mechanical mode directly depends on the classical amplitude of the TEM$_{01}$ and vice versa. For the entanglement between the two optical modes, this reaches a maximum when both modes have medium power. This can be attributable to the fact that the entanglement between these two optical modes is mediated by the mechanical mode, and both $E_\mathcal{N}^{0m}$ and $E_\mathcal{N}^{1m}$ should be large to give a reasonable $E_\mathcal{N}^{01}$. Besides, as shown explicitly in Fig. 6.4, the condition $\omega_1 - \omega_0 = \omega_m$ will naturally optimize the entanglement between the TEM$_{01}$ mode and the mechanical mode. This is because the Lorentzian profiles of the TEM$_{01}$ mode and the mechanical have the largest overlap when $\Delta = \omega_m$. In this case, both the TEM$_{01}$ mode and the mechanical mode are driven by the same vacuum field, which gives the maximal entanglement. Therefore, the optimal condition for the cooling experiment will simultaneously optimize the entanglement strength, as has also been observed by Genes et al. [10].

To illustrate the robustness of this tripartite entanglement, we show the dependence of $E_\mathcal{N}$ on the environmental temperature in Fig. 6.5. The entanglement between the optical modes and the mechanical mode is very robust and it persists even when the temperature goes up to 80 K. Although the entanglement between the two optical modes is relatively weak, it changes more slowly as the temperature increases, and it vanishes when the temperature becomes higher than 15 K. The robustness of the optomechanical entanglement was also shown previously by Vitali et al. [40]. This is attributable to the strong optomechanical coupling, which suppresses the thermal decoherence of the mechanical mode. With both the TEM$_{00}$ mode and the TEM$_{01}$ mode on resonance, we can obtain much higher intra-cavity power, as compared with the equivalent detuned two-mode system. Given moderate input optical power, this allows us to achieve stronger entanglement between the optical modes and the

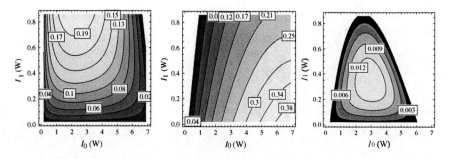

Fig. 6.3 Logarithmic negativity $E_\mathcal{N}$ as a function of the input optical powers of both modes. Other specifications are identical to those for cooling experiments given in the previous section. The *left panel* shows $E_\mathcal{N}^{0m}$ for the entanglement between the TEM$_{00}$ mode and the mechanical mode; The *middle panel* presents $E_\mathcal{N}^{1m}$ for the TEM$_{01}$ mode and the mechanical mode; the *right panel* shows $E_\mathcal{N}^{01}$ for the TEM$_{00}$ mode and the TEM$_{01}$ mode. From Ref. [22]

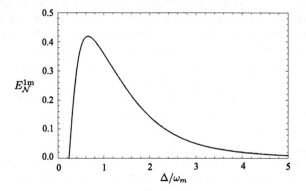

Fig. 6.4 Logarithmic negativity $E_\mathcal{N}^{1m}$ as a function of the cavity modes gap $\Delta \equiv \omega_1 - \omega_0$. As we can see, the condition $\Delta \approx \omega_m$, which optimizes the cooling, also maximizes the entanglement between the TEM$_{01}$ mode and the mechanical mode. Here we have assumed $I_0 = 4.5$ W (higher I_0 will make the system unstable for small Δ) and $I_1 = 0$. W. Since it can be viewed as an effective two-mode system in this case with $I_1 = 0$, we simply recover the results given by Vitali et al. [19]. From Ref. [22]

mechanical mode of a massive mechanical oscillator ($\sim$ mg). Of course, this robustness of entanglement is conditional on the fact that the mirrors of the cavity can sustain a high optical power $\sim 10^4$ W. If the beam size is of the order of mm, this corresponds to a power density of around 10^6 W/cm^2, which is achievable with the present technology [32].

To verify this tripartite entanglement experimentally, we can apply the same protocol as proposed in Refs. [18, 28, 40]. Specifically, through measuring the outgoing field, we can build up statistics and construct the covariance matrix $\mathbf{V}_{\text{exp}}$ of this tripartite system based on the measurement results, and then analyze whether the partially-transposed covariance matrix $\mathbf{V}_{\text{exp}}^{\text{pt}}$ fails to be positive definite. If $\mathbf{V}_{\text{exp}}^{\text{pt}}$ has a negative eigenvalue, this will give an unambiguous signature for quantum

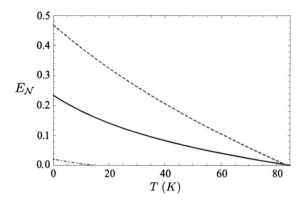

Fig. 6.5 Logarithmic negativity as a function of temperature. The *solid* curve stands for $E_{\mathcal{N}}^{0m}$, the *dashed* curve for $E_{\mathcal{N}}^{0m}$ and *dash-dot* curve for $E_{\mathcal{N}}^{01}$. We have chosen the optimal parameters for each curve. From Ref. [32]

entanglement, because any classical correlation always gives a positive definite $\mathbf{V}_{\text{exp}}^{\text{pt}}$. Besides, we can also use $\mathbf{V}_{\text{exp}}$ to evaluate the logarithmic negativity $E_{\mathcal{N}}$ of any bipartite subsystem to determine whether entanglement exists or not in a given subsystem. Since the tripartite entanglement is stationary, this means that the optomechanical interactions protect the quantum entanglement from the thermal decoherence, which is a significant problem in non-stationary quantum entanglements. In principle, we can make a sufficiently long integration of the output signal such that the shot noise is negligibly small, and $\mathbf{V}_{\text{exp}}$ should be a direct verification of what we have obtained theoretically.

6.6 Three-Mode Interactions With a Coupled Cavity

In this section, we will discuss how to explore three-mode interactions using a coupled cavity. To make our analysis close to realistic experiments, we consider a torsional acoustic mode with frequency ~ 1 MHz interacting with the optical TEM_{10} and TEM_{00} modes. This configuration is chosen because MHz frequency can be easily achieved in a mm-scale structure, and the torsional mode has a large spatial overlap with the TEM_{10} mode.

To begin with, let us consider a single Fabry-Pérot cavity to see why a coupled cavity is necessary. The free spectral range of a single cavity with length ~ 10 cm is approximately 1 GHz. Therefore, as shown in Fig. 6.6, one has to build either a near-planar or near-concentric cavity to obtain a desired mode gap around 1 MHz between the TEM_{10} and TEM_{00} modes. For both cases, the cavity is marginally stable and susceptible to misalignment. It is also difficult to get accesses to both instability and cooling regimes in a single setup. Creating a coupled Fabry-Perot cavity solves these problems. As we will show later, the resulting scheme is stable. Additionally, we can easily tune between the instability and cooling regimes. The coupled cavity is showed schematically in Fig. 6.7. It is similar to the configuration of power- or signal-recycling interferometers [6, 21] when one considers either the common mode

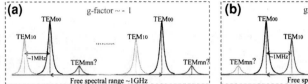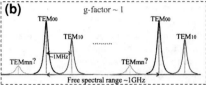

Fig. 6.6 The optical modes of a single Fabry-Pérot cavity with length ~10 cm. The panel (**a**) shows the mode distribution for a near-concentric cavity with g-factor ~−1 which is suitable for observing PI, while panel (**b**) is the near-planar case with g-factor ~1 which suits for cooling experiment. In both cases, there are no symmetric modes on the opposite side of the TEM_{00} mode because the higher-order mode TEM_{mn} marked with '?' are highly lossy due to diffraction losses. This is preferred for experimental realizations of three-mode interactions because we know from Eq. (6.8) that any symmetric mode on the opposite side of the TEM_{00} mode will reduce the absolute value of the parametric gain. However, both cavities are marginally stable and very susceptible to misalignment. From Ref. [42]

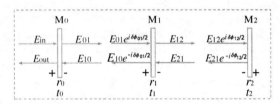

Fig. 6.7 The optical fields of the coupled cavity. Here $\delta\phi_{01,12}$ are the round-trip phase shifts of light in the sub-cavity (formed by M_0 and M_1,) and in the main cavity (formed by M_1 and M_2,) respectively. We use the convention that the mirrors have minus reflectivity on the side with a coating layer. From Ref. [42]

or the differential mode. The field dynamics can be easily obtained as shown in Ref. [31], by treating the sub-cavity as an effective mirror, with frequency- and mode-dependent transmissivity and reflectivity. Specifically, the effective transmissivity t_{01} is:

$$t_{01} \equiv \frac{E_{12}}{E_{in}} = \frac{t_0 t_1}{1 + r_0 r_1 e^{i\delta\phi_{01}}}, \quad (6.54)$$

and the effective reflectivity r_{10} is given by:

$$r_{10} \equiv \frac{E_{12}}{E_{21}} = -r_1 - \frac{t_1^2 r_0 e^{i\delta\phi_{01}}}{1 + r_0 r_1 e^{i\delta\phi_{01}}}. \quad (6.55)$$

The corresponding E_{12} inside the main cavity can be written as:

$$E_{12} = \frac{E_{in} t_{01}}{1 + r_{10} r_2 e^{i\delta\phi_{12}}} = \frac{E_{in} t_{01}}{1 - |r_{10}| r_2 e^{i[\arg(r_{10}) + \delta\phi_{12} + \pi]}}. \quad (6.56)$$

The resonance occurs when the phase factor in Eq. (6.56) is equal to $2n\pi$, which critically depends upon the phase angle of the effective reflectivity, namely $\arg(r_{10})$.

6.6 Three-Mode Interactions With a Coupled Cavity

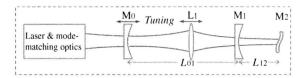

Fig. 6.8 The optical layout for the table top experiment, where a 1 MHz torsional micro-oscillator (M_2) interacts with the optical TEM_{10} mode and the TEM_{00} mode. By tuning the positions of mirror M_0 and lens L_1, we can continuously change the frequency of TEM_{10}. If the losses in L_1 were an issue, it could easily be replaced by a concave mirror. From Ref. [42]

Specifically, when the TEM_{00} mode resonates inside the main cavity, which requires that $\delta\phi_{01}^{TEM_{00}} = \delta\phi_{12}^{TEM_{00}} = 2n\pi$, the phase shift of the TEM_{10} mode $\delta\phi_{ij}^{TEM_{10}}$ is:

$$\delta\phi_{ij}^{TEM_{10}} = \frac{2L_{ij}}{c}\Delta\omega - 2\Phi_g^{ij} + 2n'\pi, \quad ij = 01, 12, \tag{6.57}$$

where $\Delta\omega \equiv \omega_1 - \omega_0$ is the mode gap between TEM_{10} and TEM_{00}; Φ_g is the Gouy phase and n, n' are integers. In order to satisfy the resonant condition for three-mode interactions, we need to adjust $\delta\phi_{01}^{TEM_{10}}$, which changes $\arg(r_{10})$, such that $\Delta\omega = \pm\omega_m$. To achieve this, one obvious way is to change the length of sub-cavity L_{01} but this turns out to be impractical due to a small tuning range. An alternative and more practical approach, as shown in Fig. 6.8, is to add another lens or concave mirror inside the sub-cavity to tune the Gouy phase Φ_g^{01}. The resulting scheme is similar to the proposed stable recycling cavity for next-generation gravitational-wave detectors [25]. With the additional lens, the Gaussian beam gets focused inside the sub-cavity. Since the Gouy phase changes from almost $-\frac{\pi}{2}$ to $\frac{\pi}{2}$ within one Rayleigh range around the waist, one can easily obtain a desired $\delta\phi_{01}^{TEM_{10}}$ simply by adjusting the position of M_0 near the waist. This might lead to problems with power density due to the small waist size, but for the table top experiment we consider here, the power density is quite low.

The corresponding Φ_g^{01} with an additional lens can be derived straightforwardly by using the ray transfer relation for a Gaussian beam, which is given by

$$q' = \frac{fq}{f - q}. \tag{6.58}$$

Here f is the focal length of L_1; $q^{(\prime)} \equiv z^{(\prime)} + iz_R^{(\prime)}$; z is the displacement relative to the waist; z_R is the Rayleigh range and superscript $'$ denotes quantities after the lens. This dictates

$$z' = \frac{f(zf - z^2 - z_R^2)}{(f-z)^2 + z_R^2}, \tag{6.59}$$

$$z_R' = \frac{z_R f^2}{(f-z)^2 + z_R^2}. \tag{6.60}$$

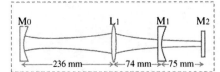 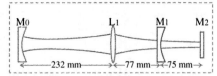

(a) Mode matching for Stokes mode (R>0) **(b)** Mode matching for Anti-Stokes mode (R<0)

Fig. 6.9 The mode matching for the positive gain and negative gain by adjusting the relative position of M_0 and L_1. Only small adjustment is needed to tune from one case to another. From Ref. [42]

The resulting Gouy phase at any point is given by

$$\Phi_g(z) = \begin{cases} \arctan(z/z_R), & z < z_L \\ \arctan\left[(z - z_L + z'_L)/z'_R\right] + \arctan(z_L/z_R) - \arctan(z'_L/z'_R), & z \geq z_L \end{cases} \quad (6.61)$$

where $z_L^{(\prime)}$ is the position of L_1 relative to the waist. Gouy phase Φ_g^{01} is the difference between wavefront at M_0 and M_1, namely

$$\Phi_g^{01} = \Phi_g(z_{M_0}) - \Phi_g(z_{M_1}), \quad (6.62)$$

where z_{M_0} and z_{M_1} are the positions of M_0 and M_1 relative to the waist respectively. Therefore, by adjusting the positions of M_0 and L_1 as shown in Fig. 6.8, we can continuously tune Φ_g^{01} such that $\Delta\omega = \omega_m$.

Equations (6.54)–(6.62) provide the design tools of the coupled cavity for three-mode interactions. To realize the experiment, we first need to design the main cavity and specify L_{12}, ω_m, the radius of curvatures (RoCs) of M_0, M_1, M_2 and the focal length of L_1. From Eqs. (6.56) and (6.57), we can find out the required $\arg(r_{01})$ which gives the right mode gap between TEM_{10} and TEM_{00}. This will gives us one constraint. Combining with the requirement of mode matching to M_0, we can fix two degrees of freedom of the system, namely the positions of M_0 and L_1. To demonstrate this principle explicitly, we present a solution that is close to a realistic experimental setup. We assume the following:

$$L_{12} = 75\,\text{mm} \quad \omega_m = 1\,\text{MHz} \quad f = 100\,\text{mm},$$
$$R_0 = 500\,\text{mm} \quad r_0 = \sqrt{0.999} \quad A_0 = 500\,\text{ppm},$$
$$R_1 = 100\,\text{mm} \quad r_1 = \sqrt{0.9} \quad A_1 = 500\,\text{ppm},$$
$$R_2 = \infty\,\text{mm} \quad r_2 = \sqrt{0.9995} \quad A_2 = 500\,\text{ppm}.$$

Here, $R_i(i = 0, 1, 2)$ are RoCs; r_i denotes the amplitude reflectivity; t_i is the amplitude transmissivity and A_i is the optical loss which satisfy $r_i^2 + t_i^2 + A_i = 1$ ($i = 0, 1, 2$).

The results of mode matching for both positive- (instability) and negative-gain (cooling) configurations are shown in Fig. 6.9. In Fig. 6.10, we show the mode gap between TEM_{10} and TEM_{00} as a function of the position of M_0 relative to L_1 and the position of L_1 relative to M_1. In this particular case, the dependence is almost

6.6 Three-Mode Interactions With a Coupled Cavity

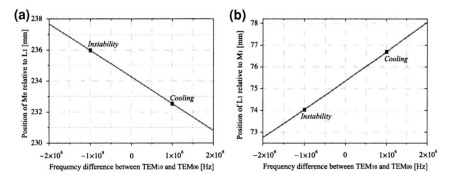

Fig. 6.10 Panels (**a**) and (**b**) show the mode gap between TEM_{10} and TEM_{00} as a function of position of M_0 relative to L_1 and position of L_1 relative to M_1 respectively. The *dots* in both figures are the situations considered in Fig. 6.9. Clearly, we can tune between the instability and cooling regimes continuously. From Ref. [42]

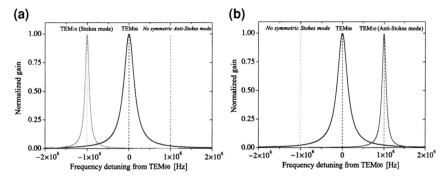

Fig. 6.11 The normalized gain of the TEM_{00} mode and the TEM_{10} mode in the positive and negative gain configurations. The mode gap is equal to $\omega_m \sim 1$ MHz, which fulfils the resonant condition for the three-mode opto-acoustic interactions. Here we simply assume that the size of the mirrors is infinite so the quality factor of the TEM_{10} mode is solely due to optical losses such as absorption. This assumption is reasonable when the mode number is small. Given the specifications in the main text, $Q_a \approx Q_s = 2.4 \times 10^9$ and $\omega_m/\gamma_a < 1$. Therefore, it can be implemented in the resolved-sideband cooling. (**a**) the TEM_{10} mode is 1 MHz below the TEM_{00} mode; (**b**) the TEM_{10} mode is 1 MHz above the TEM_{00} mode. From Ref. [42]

linear with a slope ~ 2 mm/MHz for both panels. This indicates that to tune within a cavity linewidth ~ 0.1 MHz, the mirror position needs to be adjusted within several 100 μm, which can be achieved easily. Therefore, we can continuously tune between instability and cooling regimes. Figure 6.11 shows the resulting resonance curves for both cases with the corresponding mode matching shown in Fig. 6.9. The corresponding mode gap between the TEM_{10} and TEM_{00} modes is equal to $\omega_m \sim 1$ MHz. More importantly, there is no symmetric mode on the opposite side of the TEM_{00} mode, whose presence could contribute a parametric gain with the opposite sign, thereby suppressing the overall effects. The absolute value of parametric gain $\mathcal{R}$ could be larger than 1, if we further assume that the intra-cavity power I_c is 100 mW, $Q_m = 10^6$, the mass of the oscillator $m = 1$ mg and the wavelength of light is 1,064 nm. Since the cavity is in the resolved-sideband regime where the cavity linewidth is much smaller than the mechanical frequency [20], this configuration can also be applied in the resolved-sideband cooling of acoustic modes, which is less susceptible to quantum noise.

6.7 Conclusions

We have analyzed the three-mode optomechanical parametric interactions in the quantum picture. We have derived the quantum limit for cooling experiments with three-mode interactions based upon the Fluctuation-Dissipation-Theorem. We have shown the existence of tripartite quantum entanglements in this system. The simultaneous resonances of the carrier and sideband modes in the three-mode system allows more efficient mechanical-mode cooling and more robust optomechanical entanglement than in the two-mode system. This work provides the theoretical basis for the feasibility of realizing both ground-state cooling and stationary optomechanical quantum entanglements using three-mode optomechanical parametric interactions in small-scale table-top experiments and also large-scale GW interferometers.

References

1. G. Adesso, F. Illuminati, Entanglement in continuous-variable systems: recent advances and current perspectives. J. Phys. A: Math. Theor. **40**(28), 7821 (2007)
2. O. Arcizet, P.-F. Cohadon, T. Briant, M. Pinard, A. Heidmann, Radiation-pressure cooling and optomechanical instability of a micromirror. Nature **444**, 71–74 (2006)
3. V.B. Braginsky, S.E. Strigin, S.P. Vyatchanin, Parametric oscillatory instability in fabry-perot interferometer. Phys. Lett. A **287**, 331–338 (2001)
4. V.B. Braginsky, S.E. Strigin, S.P. Vyatchanin, Analysis of parametric oscillatory instability in power recycled LIGO interferometer. Phys. Lett. A **305**, 111–124 (2002)
5. V.B. Braginsky, S.P. Vyatchanin, Low quantum noise tranquilizer for fabry-perot interferometer. Phys. Lett. A **293**, 228–234 (2002)
6. A. Buonanno, Y. Chen, Quantum noise in second generation, signalrecycled laser interferometric gravitational-wave detectors. Phys. Rev. D **64**, 042006 (2001)
7. P.F. Cohadon, A. Heidmann, M. Pinard, Cooling of a mirror by radiation pressure. Phys. Rev. Lett. **83**, 3174 (1999)
8. T. Corbitt, Y. Chen, E. Innerhofer, H. Mueller-Ebhardt, D. Ottaway, H. Rehbein, D. Sigg, S. Whitcomb, C. Wipf, N. Mavalvala, An all-optical trap for a gram-scale mirror. Phys. Rev. Lett. **98**, 150802–150804 (2007)
9. T. Corbitt, C. Wipf, T. Bodiya, D. Ottaway, D. Sigg, N. Smith, S. Whitcomb, N. Mavalvala, Optical dilution and feedback cooling of a gram-scale oscillator to 6.9 mK . Phys. Rev. Lett. **99**, 160801–160804 (2007)
10. C. Genes, D. Vitali, P. Tombesi, Simultaneous cooling and entanglement of mechanical modes of a micromirror in an optical cavity. New J. Phys. **10**(9), 095009 (2008)

11. C. Genes, D. Vitali, P. Tombesi, S. Gigan, M. Aspelmeyer, Groundstate cooling of a micromechanical oscillator: comparing cold damping and cavity-assisted cooling schemes. Phys. Rev. A **77**, 033804 (2008)
12. S. Gigan, H.R. Böhm, M. Paternostro, F. Blaser, G. Langer, J.B. Hertzberg, K.C. Schwab, D. Bäuerle, M. Aspelmeyer, A. Zeilinger, Self-cooling of a micromirror by radiation pressure. Nature **444**, 67–70 (2006)
13. S. Gras, D. Blair, L. Ju, Test mass ring dampers with minimum thermal noise. Phys. Lett. A **372**, 1348–1356 (2008)
14. A. Gurkovsky, S. Strigin, S. Vyatchanin, Analysis of parametric oscillatory instability in signal recycled LIGO interferometer. Phys. Lett. A **362**, 91–99 (2007)
15. M.J. Hartmann, M.B. Plenio, Steady state entanglement in the mechanical vibrations of two dielectric membranes. Phys. Rev. Lett., **101**, 200503 (2008)
16. L. Ju, S. Gras, C. Zhao, J. Degallaix, D. Blair, Multiple modes contributions to parametric instabilities in advanced laser interferometer gravitational wave detectors. Phys. Lett. A **354**, 360–365 (2006) June
17. D. Kleckner, D. Bouwmeester, Subkelvin optical cooling of a micromechanical resonator. Nature **444**, 75–78 (2006)
18. J. Laurat, G. Keller, J. Augusto, J.A. Oliveira-Huguenin, C. Fabre, T. Coudreau, A. Serafini, G. Adesso, F. Illuminati, Entanglement of two-mode Gaussian states: characterization and experimental production and manipulation. J. Opt. B: Quantum and Semiclassical Opt. **7**(12), S577 (2005)
19. S. Mancini, D. Vitali, P. Tombesi, Scheme for teleportation of quantum states onto a mechanical resonator. Phys. Rev. Lett. **90**, 137901 (2003)
20. F. Marquardt, J.P. Chen, A.A. Clerk, S.M. Girvin, Quantum theory of cavity-assisted sideband cooling of mechanical motion. Phys. Rev. Lett. **99**, 093902 (2007)
21. B.J. Meers, Recycling in laser-interferometric gravitational-wave detectors. Phys. Rev. D **38**, 2317 (1988)
22. H. Miao, C. Zhao, L. Ju, D.G. Blair, Quantum ground-state cooling and tripartite entanglement with three-mode optoacoustic interactions. Phys. Rev. A **79**, 063801 (2009)
23. H. Miao, C. Zhao, L. Ju, S. Gras, P. Barriga, Z. Zhang, G. Blair D., Three-mode optoacoustic parametric interactions with a coupled cavity. Phys. Rev. A **78**, 063809 (2008)
24. C.M. Mow-Lowry, A.J. Mullavey, S. Gossler, M.B. Gray, D.E. Mc-Clelland, Cooling of a gram-scale cantilever flexure to 70 mK with a servo-modified optical spring. Phys. Rev. Lett. **100**, 010801–010804 (2008)
25. G. Mueller, LIGO document: G070441-00-R (2007)
26. H. Mueller-Ebhardt, H. Rehbein, R. Schnabel, K. Danzmann, Y. Chen, Entanglement of macroscopic test masses and the standard quantum limit in laser interferometry. Phys. Rev. Lett. **100**, 013601 (2008)
27. A. Naik, O. Buu, M.D. LaHaye, A.D. Armour, A.A. Clerk, M.P. Blencowe, K.C. Schwab, Cooling a nanomechanical resonator with quantum backaction. Nature **443**, 193–196 (2006)
28. M. Paternostro, D. Vitali, S. Gigan, M.S. Kim, C. Brukner, J. Eisert, M. Aspelmeyer, Creating and probing multipartite macroscopic entanglement with light. Phys. Rev. Lett. **99**, 250401 (2007)
29. A. Peres, Separability criterion for density matrices. Phys. Rev. Lett. **77**, 1413 (1996)
30. M. Poggio, C.L. Degen, H.J. Mamin, D. Rugar, Feedback cooling of a cantilever's fundamental mode below 5 mK. Phys. Rev. Lett. **99**, 017201–017204 (2007)
31. M. Rakhmanov, Dynamics of laser interferometric gravitational wave detectors. PhD thesis, California Institute of Technology (2003)
32. D. Reitze, LIGO R & D documents: LIGO-G050360-00-R (2005)
33. S.W. Schediwy, C. Zhao, L. Ju, D.G. Blair, P. Willems, Observation of enhanced optical spring damping in a macroscopic mechanical resonator and application for parametric instability control in advanced gravitationalwave detectors. Phys. Rev. A **77**, 013813–013815 (2008)

34. A. Schliesser, P. Del'Haye, N. Nooshi, K.J. Vahala, T.J. Kippenberg, Radiation pressure cooling of a micromechanical oscillator using dynamical backaction. Phys. Rev. Lett. **97**, 243905-4 (2006)
35. A. Schliesser, R. Riviere, G. Anetsberger, O. Arcizet, T.J. Kippenberg, Resolved-sideband cooling of a micromechanical oscillator. Nat. Phys. **4**, 415–419 (2008)
36. A. Serafini, Multimode uncertainty relations and separability of continuous variable states. Phys. Rev. Lett. **96**, 110402 (2006)
37. R. Simon, Peres-Horodecki separability criterion for continuous variable systems. Phys. Rev. Lett. **84**, 2726 (2000) Mar
38. J.D. Thompson, B.M. Zwickl, A.M. Jayich, F. Marquardt, S.M. Girvin, J.G.E. Harris, Strong dispersive coupling of a high-finesse cavity to a micromechanical membrane. Nature **452**, 72–75 (2008)
39. G. Vidal, R.F. Werner, Computable measure of entanglement. Phys. Rev. A **65**, 032314 (2002)
40. D. Vitali, S. Gigan, A. Ferreira, H.R. Bohm, P. Tombesi, A. Guerreiro, V. Vedral, A. Zeilinger, M. Aspelmeyer, Optomechanical entanglement between a movable mirror and a cavity field. Phys. Rev. Lett. **98**, 030405 (2007)
41. I. Wilson-Rae, N. Nooshi, W. Zwerger, T.J. Kippenberg, Theory of ground state cooling of a mechanical oscillator using dynamical backaction. Phys. Rev. Lett. **99**, 093901 (2007)
42. C. Zhao, L. Ju, J. Degallaix, S. Gras, D.G. Blair, Parametric instabilities and their control in advanced interferometer gravitational-wave detectors. Phys. Rev. Lett. **94**, 121102 (2005)
43. C. Zhao, L. Ju, Y. Fan, S. Gras, B.J.J. Slagmolen, H. Miao, P. Barriga, D.G. Blair, D.J. Hosken, A.F. Brooks, P.J. Veitch, D. Mudge, J. Munch, Observation of three-mode parametric interactions in long optical cavities. Phys. Rev. A **78**, 023807 (2008)
44. C. Zhao, L. Ju, H. Miao, S. Gras, Y. Fan, D.G. Blair, Three-mode optoacoustic parametric amplifier: a tool for macroscopic quantum experiments. Phys. Rev. Lett. **102**, 243902 (2009)

Chapter 7
Achieving the Ground State and Enhancing Optomechanical Entanglement

7.1 Preface

In the previous chapter, we have seen that in order to achieve the quantum ground state of a mechanical oscillator, with three-mode or even general optomechanical devices, the cavity bandwidth needs to be smaller than the mechanical frequency. This is the so-called resolved-sideband or good-cavity limit. In this chapter, we provide a new but physically equivalent insight into the origin of such a limit: that is information loss due to a finite cavity bandwidth. With an optimal feedback control to recover this information, we can surpass the resolved-sideband limit and achieve the quantum ground state. Interestingly, recovering this information can also significantly enhance the optomechanical entanglement. Especially when the environmental temperature is high, the entanglement will either exist or vanish, depending critically on whether the information is recovered or not, which is a vivid example of a quantum eraser. This is a joint research effort by Stefan Danilishin, Helge Müller-Ebhardt, Yanbei Chen, and myself.

7.2 Introduction

Recently, achieving the quantum ground state of a macroscopic mechanical oscillator has triggered great interest among physicists. It will not only have significant impact on quantum-limited measurements [5] but will also shed light on quantum entanglements involving macroscopic mechanical degrees of freedom [4, 24, 32, 40, 44, 57], which can be useful for future quantum computing and help us to understand transitions between the classical and quantum domains [12, 13, 45, 6].

By using conventional cryogenic refrigeration, O'Connell et al. have successfully cooled a 6 GHz micromechanical oscillator down to its ground state [43]. Meanwhile, in order to cool larger-size and lower-frequency mechanical oscillators at high environmental temperature, there have been great efforts in trying different approaches:

active feedback control, and parametrically coupling the oscillator to optical or electrical degrees of freedom [2, 3, 7–10, 18, 21, 23, 27, 30, 35, 39, 42, 47, 49, 50, 51, 54, 55]. The cooling mechanism has been extensively discussed, and certain classical and quantum limits have been derived [11, 14, 20, 31, 33, 48, 58, 60, 59]. In the case of cavity-assisted cooling schemes, pioneering theoretical works by Marquardt et al. [33] and Wilson-Rae et al. [59] showed that the quantum limit for the occupation number is $(\gamma/2\,\omega_m)^2$.[1] This dictates that, in order to achieve the ground state of the mechanical oscillator, the cavity bandwidth γ must be smaller than the mechanical frequency ω_m, which is the so-called "resolved-sideband" or "good-cavity" limit. This limit is derived by analyzing the quantum fluctuations of the radiation pressure force on the mechanical oscillator. From a physically equivalent perspective, it can actually be attributable to information loss: information of the oscillator motion leaks into the environment without being carefully treated, which induces decoherence.

This perspective immediately illuminates two possible approaches for surpassing such a limit: (i) the *first* one is to implement the novel scheme proposed by Elste et al. [17], in which the quantum noise is destructively interfered, and information of the oscillator motion around ω_m does not leak into the environment. Corbitt suggested an intuitive understanding by thinking of an optical cavity with a movable front mirror rather than a movable end mirror in those cooling experiments (private communication). In this hypothetical scheme, the optical fields directly reflected, and those filtered through the cavity, both contain the information of the front-mirror motion. If the cavity detuning is appropriate, these two pieces of information destructively interfere with each other, and the quantum coherence of the mechanical oscillator is maintained. (ii) the *second* approach is to recover the information by detecting the cavity output. This will work because a conditional quantum state—*the best knowledge of the oscillator state, conditional on the measurement result*—is always pure for an ideal continuous measurement with no readout loss. Indeed, when the cavity bandwidth is much larger than the mechanical frequency, the cavity mode will follow the oscillator dynamics and can therefore be adiabatically eliminated. The quantum noise can be treated as being Markovian and a standard Stochastic-Master-Equation (SME) analysis has already shown how the conditional quantum state approaches a pure state under a continuous measurement [15, 16, 19, 25, 38]. For the non-zero cavity bandwidth considered here, the cavity mode has a dynamical timescale comparable to that of the mechanical oscillator. Correspondingly, the quantum noise has correlations at different times, and is non-Markovian. To estimate the conditional state, a Wiener-filtering approach is more transparent than the SME [14]. As we will show, the conditional quantum state of the oscillator in the cavity-assisted cooling schemes is indeed almost pure, with a residual impurity contributed by the thermal noise, and by imperfections in detections and optomechanical entanglement between the oscillator and the cavity mode. In order to further localize the oscillator in phase space and achieve its ground state, an optimal feedback control is essential [11].

---

[1] There is a factor of two difference in defining the cavity bandwidth here compared with the one defined in Ref. [33].

7.2 Introduction

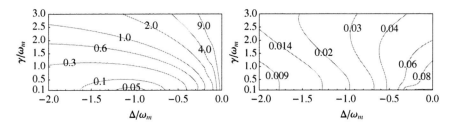

Fig. 7.1 A contour plot of the occupation number as a function of cavity bandwidth γ and detuning Δ for the unconditional state (*left*) as obtained in Refs. [33, 39] and optimally controlled state (*right*), for which the details are in Sects. 7.4, 7.5, 7.6. From Ref. [50]

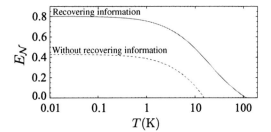

Fig. 7.2 Optomechanical entanglement strength $E_\mathcal{N}$ as a function of temperature T with (*solid*) and without (*dashed*) recovering information (details are in Sect. 7.7). From Ref. [37]

In Fig. 7.1, the final occupation number of the unconditional state and optimally controlled state is shown. As long as the optimal control is applied, the minimally achievable occupation number of the oscillator will not be constrained by the resolved-sideband limit.

Another interesting issue in the optomechanical system is creating a quantum entanglement between the cavity mode and the oscillator, or even between two oscillators [24, 31, 34, 40, 44, 57]. Intuitively, one might think that such an entanglement must be very vulnerable to thermal decoherence, and that the environmental temperature needs to be extremely low in order to create it. However, as shown in Ref. [40], and in a more recent investigation [36], the environmental temperature—even though being an important factor—affects the entanglement implicitly, and only the ratio between the interaction strength and thermal decoherence matters. The reason why, in Refs. [24, 31, 44, 57], the temperature plays a dominant role in determining the existence of the entanglement can also be traced back to information loss, as briefly mentioned in Ref. [36]. Here, we will address this issue more explicitly. Figure 7.2 shows that by recovering the information contained in the cavity output, the optomechanical entanglement can even be revived at high temperature. This is a vivid example of a quantum eraser first proposed by Scully and Drühl [52] and later demonstrated experimentally [28]: quantum coherence can be revived by recovering lost information.

The outline of this chapter is as follows: in Sect. 7.3, we will analyze the system dynamics by applying the standard Langevin-equation approach and derive the spectral densities of important dynamical quantities. In Sect. 7.4, we obtain unconditional

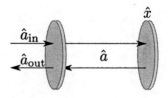

Fig. 7.3 Schematic plot of an optomechanical system with a mechanical oscillator $\hat{x}$ coupled to a cavity mode $\hat{a}$, which in turn couples to the external ingoing ($\hat{a}_{\text{in}}$) and outgoing ($\hat{a}_{\text{out}}$) optical field. From Ref.[37]

variances of the oscillator position and momentum, and evaluate the corresponding occupation number, which recovers the resolved-sideband limit. In Sect. 7.5, conditional variances are derived via the Wiener-filtering approach, which clearly demonstrates that the conditional quantum state is almost pure. In Sect. 7.6, we show the occupation number of the optimally controlled state and the corresponding optimal controller to achieve it. In Sect. 7.7, we consider the optomechanical entanglement and demonstrate that significant enhancements in the entanglement strength can be achieved after recovering information. In Sect. 7.8, to motivate cavity-assisted cooling experiments, we consider imperfections in a real experiment and obtain a numerical estimation of the occupation number given a set of experimentally achievable specifications. Finally, we conclude with our main results in Sect. 7.9

7.3 Dynamics and Spectral Densities

In this section, we will analyze the optomechanical dynamics and derive the spectral densities of relevant quantities which are essential for calculating the occupation number of the mechanical oscillator.

7.3.1 Dynamics

Even though the dynamics of such a system has been discussed extensively in the literature [20, 33, 59], we will go through some equations for the coherence of this book. An optomechanical system and the relevant dynamical quantities are shown schematically in Fig. 7.3. The corresponding Hamiltonian is given by:

$$\hat{\mathcal{H}} = \hbar \omega_c \hat{a}^\dagger \hat{a} + \frac{\hat{p}^2}{2m} + \frac{1}{2} m \omega_m^2 \hat{x}^2 + \hbar G_0 \hat{x} \, \hat{a}^\dagger \hat{a} + i \hbar \sqrt{2\gamma} \, (\hat{a}_{\text{in}} e^{-i\omega_0 t} \hat{a}^\dagger - \text{H.c.}). \quad (7.1)$$

Here, ω_c and ω_0 are the cavity resonant frequency and the laser frequency, respectively; $\hat{a}$ is the annihilation operator for the cavity mode, which satisfies $[\hat{a}, \hat{a}^\dagger] = 1$; $\hat{x}$ and $\hat{p}$ denote the oscillator position and momentum, with $[\hat{x}, \hat{p}] = i\hbar$; m is the mass of the oscillator; $G_0 \equiv \omega_0/L$ is the optomechanical coupling constant, with L the cavity length. In the rotating frame at the laser frequency ω_0, a set of nonlinear Langevin equations can be obtained:

$$\dot{\hat{x}}(t) = \hat{p}(t)/m, \quad (7.2)$$

7.3 Dynamics and Spectral Densities

$$\dot{\hat{p}}(t) = -\gamma_m \hat{p}(t) - m\,\omega_m^2 \hat{x}(t) - \hbar\,G_0 \hat{a}^\dagger(t)\hat{a}(t) + \hat{\xi}_{\text{th}}(t), \qquad (7.3)$$

$$\dot{\hat{a}}(t) = -(\gamma - i\Delta)\hat{a}(t) - i\,G_0 \hat{x}(t)\hat{a}(t) + \sqrt{2\gamma}\,\hat{a}_{\text{in}}(t), \qquad (7.4)$$

where the cavity detuning $\Delta \equiv \omega_0 - \omega_c$. To take into account the fluctuation-dissipation mechanism of the oscillator coupled to a thermal heat bath at temperature T, we have included the mechanical damping γ_m and the associated Brownian force $\hat{\xi}_{\text{th}}$, of which the correlation function is $\langle \hat{\xi}_{\text{th}}(t)\hat{\xi}_{\text{th}}(t')\rangle = 2m\,\gamma_m k_B T \delta(t - t')$ in the high-temperature limit. In a cooling experiment, the cavity mode is driven by a coherent laser and, to a good approximation, the system is linear. To linearize the system, we simply replace any operator $\hat{o}(t)$ with the sum of a steady-state part and a small perturbed part, namely $\hat{o}(t) \to \bar{o} + \hat{o}(t)$.[2] We assume that the mean displacement of the oscillator is equal to zero: $\bar{x} = 0$. The solution to $\bar{a}$ is simply $\bar{a} = \sqrt{2\gamma}\,\bar{a}_{\text{in}}/(\gamma - i\Delta)$, and $\bar{a}_{\text{in}} = \sqrt{I_0/(\hbar\omega_0)}$ with I_0 the input optical power. We have chosen an appropriate phase reference such that $\bar{a}$ is real and positive. The resulting linearized equations are:

$$m[\ddot{\hat{x}}(t) + \gamma_m \dot{\hat{x}}(t) + \omega_m^2 \hat{x}(t)] = -\hbar\,\bar{G}_0[\hat{a}^\dagger(t) + \hat{a}(t)] + \hat{\xi}_{\text{th}}(t), \qquad (7.5)$$

$$\dot{\hat{a}}(t) + (\gamma - i\Delta)\hat{a}(t) = -i\,\bar{G}_0 \hat{x}(t) + \sqrt{2\gamma}\,\hat{a}_{\text{in}}(t), \qquad (7.6)$$

with $\bar{G}_0 \equiv G_0 \bar{a}$. The input-output relation of the cavity, which relates the cavity mode to the external continuum optical mode, is [19]:

$$\hat{a}_{\text{out}}(t) = \sqrt{\eta}[-\hat{a}_{\text{in}}(t) + \sqrt{2\gamma}\,\hat{a}(t)] + \sqrt{1-\eta}\,\hat{n}(t), \qquad (7.7)$$

where η is the quantum efficiency of the photodetector, and $\hat{n}$ is the associated vacuum fluctuation that is not correlated with $\hat{a}_{\text{in}}$. The linearized dynamics of this system are fully described by Eqs. (7.5), (7.6) and (7.7) which can be solved in the frequency domain.

Mechanical oscillator part—By denoting the Fourier component of any quantity O as $\tilde{O}(\Omega)$, the solution for the oscillator position is:

$$\tilde{x}(\Omega) = \tilde{R}_{\text{eff}}(\Omega)[\tilde{F}_{\text{BA}}(\Omega) + \tilde{\xi}_{\text{th}}(\Omega)]. \qquad (7.8)$$

Here, the back-action force $\tilde{F}_{\text{BA}}(\Omega)$ is:

$$\tilde{F}_{\text{BA}}(\Omega) = 2\hbar\,\bar{G}_0 \sqrt{\gamma}\,\chi(\Omega)[(\gamma - i\Omega)\tilde{v}_1(\Omega) - \Delta\,\tilde{v}_2(\Omega)], \qquad (7.9)$$

where we have defined the amplitude quadrature $\tilde{v}_1(\Omega)$ and the phase quadrature $\tilde{v}_2(\Omega)$ of the vacuum fluctuation, namely $\tilde{v}_1(\Omega) \equiv [\tilde{a}_{\text{in}}(\Omega) + \tilde{a}_{\text{in}}^\dagger(-\Omega)]/\sqrt{2}$ and $\tilde{v}_2(\Omega) \equiv [\tilde{a}_{\text{in}}(\Omega) - \tilde{a}_{\text{in}}^\dagger(-\Omega)]/(i\sqrt{2})$. Due to the well-known "optical-spring" effect,

---

[2] For simplicity, we use the same $\hat{o}$ to denote its perturbed part.

the mechanical response of the oscillator is modified from its original value $\tilde{R}_{xx}(\Omega) = -[m(\Omega^2 + 2i\gamma_m\Omega - \omega_m^2)]^{-1}$ to an effective one given by:

$$\tilde{R}_{\text{eff}}(\Omega) \equiv [\tilde{R}_{xx}^{-1}(\Omega) - \tilde{\Gamma}(\Omega)]^{-1}, \qquad (7.10)$$

with $\tilde{\Gamma}(\Omega) \equiv 2\hbar \bar{G}_0^2 \Delta \chi$, and $\chi \equiv [(\Omega + \Delta + i\gamma)(\Omega - \Delta + i\gamma)]^{-1}$.

Cavity mode part—The solution for the cavity mode is:

$$\tilde{a}(\Omega) = \frac{\bar{G}_0\,\tilde{x}(\Omega) + i\sqrt{2\gamma}\,\tilde{a}_{\text{in}}(\Omega)}{\Omega + \Delta + i\gamma}. \qquad (7.11)$$

In terms of amplitude and phase quadratures, this can be rewritten as:

$$\tilde{a}_1(\Omega) = \sqrt{2\gamma}\,\chi[(-\gamma + i\Omega)\tilde{v}_1(\Omega) + \Delta\,\tilde{v}_2(\Omega)] - \sqrt{2}\,\bar{G}_0\,\chi\,\Delta\,\tilde{x}(\Omega), \qquad (7.12)$$

$$\tilde{a}_2(\Omega) = \sqrt{2\gamma}\,\chi[-\Delta\,\tilde{v}_1(\Omega) - (\gamma - i\Omega)\tilde{v}_2(\Omega)] + \sqrt{2}\,\bar{G}_0\,\chi\,(\gamma - i\Omega)\,\tilde{x}(\Omega). \qquad (7.13)$$

Cavity output part—Similarly, we introduce amplitude and phase quadratures for the cavity output: $\tilde{Y}_1(\Omega) \equiv [\tilde{a}_{\text{out}}(\Omega) + \tilde{a}_{\text{out}}^\dagger(-\Omega)]/2$, and $\tilde{Y}_2(\Omega) \equiv [\tilde{a}_{\text{out}}(\Omega) - \tilde{a}_{\text{out}}^\dagger(-\Omega)]/2$. Their solutions are:

$$\tilde{Y}_i(\Omega) = \tilde{Y}_i^{\text{vac}}(\Omega) + \sqrt{\eta}\,\tilde{R}_{Y_i F}(\Omega)\,\tilde{x}(\Omega), \quad (i = 1, 2). \qquad (7.14)$$

The vacuum parts $\tilde{Y}_i^{\text{vac}}$ of the output, which induce measurement shot noise, are the following:

$$\tilde{Y}_1^{\text{vac}}(\Omega) = \sqrt{1-\eta}\,\tilde{n}_1(\Omega) + \sqrt{\eta}\,\chi[(\Delta^2 - \gamma^2 - \Omega^2)\tilde{v}_1(\Omega) + 2\gamma\Delta\,\tilde{v}_2(\Omega)], \qquad (7.15)$$

$$\tilde{Y}_2^{\text{vac}}(\Omega) = \sqrt{1-\eta}\,\tilde{n}_2(\Omega) + \sqrt{\eta}\,\chi[-2\gamma\Delta\,\tilde{v}_1(\Omega) + (\Delta^2 - \gamma^2 - \Omega^2)\tilde{v}_2(\Omega)]. \qquad (7.16)$$

The output responses $\tilde{R}_{Y_i F}(\Omega)$ are defined as [29]:

$$\tilde{R}_{Y_1 F}(\Omega) \equiv -2\sqrt{\gamma}\,\bar{G}_0\,\Delta\,\chi, \quad \tilde{R}_{Y_2 F}(\Omega) \equiv 2\sqrt{\gamma}\,\bar{G}_0(\gamma - i\Omega)\chi. \qquad (7.17)$$

7.3.2 Spectral Densities

Given the above solutions, we can analyze the statistical properties of the dynamical quantities. We consider all noises to be Gaussian and stationary but not necessarily Markovian. Their statistical properties are fully quantified by their spectral densities. We define a symmetrized single-sided spectral density $\tilde{S}_{AB}(\Omega)$ according to the standard formula [56]:

$$2\pi\delta(\Omega - \Omega')\tilde{S}_{AB}(\Omega) = \langle A(\Omega)\tilde{B}^\dagger(\Omega')\rangle_{\text{sym}} = \langle \tilde{A}(\Omega)\tilde{B}^\dagger(\Omega') + \tilde{B}^\dagger(\Omega')\tilde{A}(\Omega)\rangle. \qquad (7.18)$$

7.3 Dynamics and Spectral Densities

For vacuum fluctuations $\hat{a}_{1,2}$, we simply have $\tilde{S}_{a_1 a_1}(\Omega) = \tilde{S}_{a_2 a_2}(\Omega) = 1$ and $\tilde{S}_{a_1 a_2}(\Omega) = 0$.

Mechanical oscillator part—The spectral density for oscillator position is (cf. Eqs. (7.8) and (7.9))

$$\tilde{S}_{xx}(\Omega) = |\tilde{R}_{\text{eff}}(\Omega)|^2 \tilde{S}_{FF}^{\text{tot}}(\Omega), \tag{7.19}$$

with total force-noise spectrum:

$$\tilde{S}_{FF}(\Omega) = 4\hbar m \Omega_q^3 \gamma |\chi|^2 (\gamma^2 + \Omega^2 + \Delta^2) + 2\hbar m \Omega_F^2, \tag{7.20}$$

where we have introduced characteristic frequencies for the optomechanical interaction $\Omega_q \equiv (\hbar \bar{G}_0^2/m)^{1/3}$, and the thermal noise $\Omega_F \equiv \sqrt{2\gamma_m k_B T/\hbar}$. The spectral density for the oscillator momentum is simply $\tilde{S}_{pp}(\Omega) = m^2 \Omega^2 \tilde{S}_{xx}(\Omega)$.

Cavity mode part—The spectral density for the cavity mode is a little complicated:

$$\mathbf{S}_{aa}(\Omega) = \mathbf{M}_0 \mathbf{M}_0^\dagger + \mathbf{M}_0 \mathbf{M}_1^\dagger + \mathbf{M}_1 \mathbf{M}_0^\dagger + \mathbf{M}_2 \tilde{S}_{xx}(\Omega). \tag{7.21}$$

Here, the elements of the matrix $\mathbf{S}_{aa}$ are denoted by $\tilde{S}_{a_i a_j}(\Omega)$ ($i, j = 1, 2$); the matrix $\mathbf{M}_0$ is:

$$\mathbf{M}_0 \equiv \sqrt{2\gamma} \, \chi \begin{bmatrix} -\gamma + i\Omega & \Delta \\ -\Delta & -\gamma + i\Omega \end{bmatrix}; \tag{7.22}$$

the matrix $\mathbf{M}_1$ is:

$$\mathbf{M}_1 \equiv 2\sqrt{2\hbar} \bar{G}_0^2 \sqrt{\gamma} |\chi|^2 \tilde{R}_{\text{eff}}(\Omega) \begin{bmatrix} -\Delta(\gamma - i\Omega) & \Delta^2 \\ (\gamma - i\Omega)^2 & -\Delta(\gamma - i\Omega) \end{bmatrix}; \tag{7.23}$$

the matrix $\mathbf{M}_2$ is:

$$\mathbf{M}_2 \equiv 2\bar{G}_0^2 |\chi|^2 \begin{bmatrix} \Delta^2 & -\Delta(\gamma + i\Omega) \\ -\Delta(\gamma - i\Omega) & \gamma^2 + \Omega^2 \end{bmatrix}. \tag{7.24}$$

The cross-correlations between the cavity mode and the output $[\mathbf{S}_{aY}]_{ij} \equiv \tilde{S}_{a_i Y_j}(\Omega)$ are given by

$$\mathbf{S}_{aY} = \mathbf{M}_0 \mathbf{M}_3^\dagger + \mathbf{M}_0 \mathbf{M}_1^\dagger + \mathbf{M}_1 \mathbf{M}_3^\dagger + \sqrt{2\gamma} \mathbf{M}_2 \tilde{S}_{xx}(\Omega), \tag{7.25}$$

with

$$\mathbf{M}_3 \equiv \begin{bmatrix} \Delta^2 - \gamma^2 - \Omega^2 & 2\gamma\Delta \\ -2\gamma\Delta & \Delta^2 - \gamma^2 - \Omega^2 \end{bmatrix}. \tag{7.26}$$

The cross-correlation between the cavity mode and the oscillator is the following:

$$\begin{bmatrix} \tilde{S}_{a_1 x}(\Omega) \\ \tilde{S}_{a_2 x}(\Omega) \end{bmatrix} = 2\hbar \bar{G}_0 \sqrt{\gamma}\, \chi^* \tilde{R}^*_{\text{eff}}(\Omega) \mathbf{M}_0 \begin{bmatrix} \gamma + i\Omega \\ -\Delta \end{bmatrix} + \sqrt{2}\bar{G}_0 \chi \begin{bmatrix} -\Delta \\ \gamma - i\Omega \end{bmatrix} \tilde{S}_{xx}(\Omega).$$
(7.27)

For the oscillator momentum, $\tilde{S}_{a_k p}(\Omega) = i\, m\, \Omega\, \tilde{S}_{a_k x}$ ($k = 1, 2$).

Cavity output part—As an important feature of the quantum noise in this optomechanical system, there is a nonvanishing correlation between the shot noise $\hat{Y}_i^{\text{vac}}$ and the quantum back-action noise $\hat{F}_{\text{BA}}$, and it has the following spectral densities (cf. Eqs. (7.9), (7.15) and (7.16)):

$$\tilde{S}_{FY_1^{\text{vac}}}(\Omega) = 2\sqrt{\hbar m \gamma \eta \Omega_q^3}\, (\gamma + i\Omega)\, \chi^*,$$
(7.28)

$$\tilde{S}_{FY_2^{\text{vac}}}(\Omega) = 2\sqrt{\hbar m \gamma \eta \Omega_q^3}\, \Delta\, \chi^*$$
(7.29)

where χ^* is the complex conjugate of χ. Correspondingly, the spectral densities for the output quadratures are:

$$\tilde{S}_{Y_i Y_j}(\Omega) = \delta_{ij} + \eta\, \tilde{R}_{Y_i F}(\Omega) \tilde{R}^{\text{eff}}_{xx}(\Omega) \tilde{S}_{FY_j^{\text{vac}}}(\Omega)$$
$$+ \eta\, [\tilde{R}_{Y_j F}(\Omega) \tilde{R}^{\text{eff}}_{xx}(\Omega) \tilde{S}_{FY_i^{\text{vac}}}(\Omega)]^* + \eta\, \tilde{R}_{Y_i F}(\Omega) \tilde{R}^*_{Y_j F}(\Omega) \tilde{S}_{xx}(\Omega).$$
(7.30)

The information of the oscillator position $\hat{x}$, contained in the output $\hat{Y}_i$, is quantified by the the cross-correlations between $\hat{x}$ and $\hat{Y}_i$, which are:

$$\tilde{S}_{xY_i}(\Omega) = \sqrt{\eta}\, \tilde{R}^{\text{eff}}_{xx}(\Omega) \tilde{S}_{FY_i^{\text{vac}}}(\Omega) + \sqrt{\eta}\, \tilde{R}^*_{Y_i F}(\Omega) \tilde{S}_{xx}(\Omega).$$
(7.31)

Similarly, for the oscillator momentum, $\tilde{S}_{pY_k}(\Omega) = -i\, m\, \Omega\, \tilde{S}_{xY_k}(\Omega)$ ($k = 1, 2$).

7.4 Unconditional Quantum State and Resolved-Sideband Limit

In the red-detuned regime ($\Delta < 0$), where the cavity-assisted cooling experiments are currently working, a delayed response of the cavity mode to the oscillator motion gives rise to a viscous damping which can significantly reduce the thermal occupation number of the oscillator, as shown schematically in Fig. 7.4. Physically, it is because the mechanical response is changed into an effective one (cf. Eq. (7.10)), while the thermal force spectrum remains the same. The ground state can be achieved when the occupation number is much smaller than one. If we neglect the information of the oscillator motion contained in the output, the resulting quantum state of the oscillator will be *unconditional* and the corresponding occupation number of the oscillator can be obtained with the following standard definition:

$$\mathcal{N} \equiv \frac{1}{\hbar \omega_m} \left(\frac{V_{pp}}{2m} + \frac{1}{2} m \omega_m^2 V_{xx} \right) - \frac{1}{2},$$
(7.32)

7.4 Unconditional Quantum State and Resolved-Sideband Limit

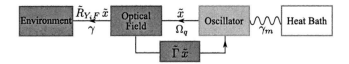

Fig. 7.4 A block diagram for an optomechanical system. The optomechanical cooling can be viewed as a feedback mechanism. This reduces the thermal occupation number of the oscillator, which has an effective temperature much lower than the heat bath. Meanwhile, some information of the oscillator motion flows into the environment without being appropriately recovered, leading to the resolved-sideband limit. From Ref. [37]

where the variances of the oscillator position V_{xx} and momentum V_{pp} are related to the spectral densities by the following formula:

$$V_{xx,pp} = \int_0^\infty \frac{d\Omega}{2\pi} \tilde{S}_{xx,pp}(\Omega). \tag{7.33}$$

Since $\mathcal{N}$ is dimensionless, it only depends on the following ratios:

$$\Omega_q/\omega_m, \ \gamma/\omega_m, \ \Delta/\omega_m, \ \Omega_F/\omega_m, \ \gamma_m/\omega_m. \tag{7.34}$$

The oscillator mass and frequency only enter implicitly. As long as these ratios are the same in different experiments, the final achievable thermal occupation number of different oscillators will be identical.

The resulting $\mathcal{N}$ is shown in the left panel of Fig. 7.1. To highlight the quantum limit, we have fixed the interaction strength Ω_q with $\Omega_q/\omega_m = 0.5$, and we have neglected the thermal force noise. In the optimal cooling regime with $\Delta = -\omega_m$, a simple closed form for the occupation number can be obtained [20]

$$\mathcal{N} = \gamma^2/(2\,\omega_m)^2 + \frac{[1 + (\gamma/\omega_m)^2](\Omega_q/\omega_m)^3}{4[1 + (\gamma/\omega_m)^2 - 2(\Omega_q/\omega_m)^3]}. \tag{7.35}$$

The resolved-sideband limit is achieved for a weak interaction strength $\Omega_q \to 0$, and

$$\mathcal{N}_{\lim} = \gamma^2/(2\,\omega_m)^2. \tag{7.36}$$

In the next section, we will demonstrate that such a limit can indeed be surpassed by recovering the information contained in the cavity output.

7.5 Conditional Quantum State and Wiener Filtering

Since, given a non-zero cavity bandwidth, the cavity output contains the information of the oscillator position (cf. Eq. (7.31)), according to the quantum mechanics, measurements of the output will collapse the oscillator wave function and project it into a *conditional* quantum state that is in accord with the measurement result. The conditional state or equivalently its Wigner function is completely determined by the conditional mean [x^{cond}, p^{cond}] and the covariance matrix $\mathbf{V}^{\text{cond}}$ between the position and momentum. More explicitly, the Wigner function reads

$$W(x, p) = \frac{1}{2\pi\sqrt{\det \mathbf{V}^{\text{cond}}}} \exp\left[-\frac{1}{2}\delta\vec{X}\,\mathbf{V}^{\text{cond}-1}\delta\vec{X}^T\right], \quad (7.37)$$

with $\delta\vec{X} = [x - x^{\text{cond}}, p - p^{\text{cond}}]$. Since more information is acquired, the conditional quantum state is always more pure than its unconditional counterpart. In the limiting case of an ideal measurement, the conditional quantum state of the mechanical oscillator would be pure with variances constrained by the *Heisenberg Uncertainty*, i.e., $\det \mathbf{V}^{\text{cond}}|_{\text{pure state}} = \hbar^2/4$.

To derive the conditional mean and variances, a mathematical tool that is usually applied is the Stochastic-Master-Equation (SME), which is most convenient for treating Markovian process [15, 16, 19, 25, 38]. In the case considered here, however, the cavity has a bandwidth comparable to the mechanical frequency, and the quantum noise is non-Markovian. The corresponding conditional mean and variance can be derived more easily with the Wiener-filtering approach. As shown in Ref. [14], the conditional mean of any quantity $\hat{o}(t)$, given a certain measurement result $Y(t')$ ($t < t'$), can be written as:

$$o(t)^{\text{cond}} \equiv \langle \hat{o}(t) \rangle^{\text{cond}} = \int_{-\infty}^{t} dt'\, K_o(t - t')Y(t'). \quad (7.38)$$

Here, $K_o(t)$ is the optimal Wiener filter, and is derived by using the standard Wiener-Hopf method. Its frequency representation is:

$$\tilde{K}_o(\Omega) = \frac{1}{\tilde{\psi}_+(\Omega)}\left[\frac{\tilde{S}_{oY}(\Omega)}{\tilde{\psi}_-(\Omega)}\right]_+ \equiv \frac{\tilde{G}_o(\Omega)}{\tilde{\psi}_+(\Omega)}, \quad (7.39)$$

where $[\,]_+$ means taking the causal component and $\tilde{\psi}_\pm$ is a spectral factorization of the output $\tilde{S}_{YY} \equiv \tilde{\psi}_+\tilde{\psi}_-$, with $\tilde{\psi}_+$ ($\tilde{\psi}_+$) and its inverse analytical in the upper-half (lower-half) complex plane, and we have introduced $\tilde{G}_o(\Omega)$. The conditional covariance between $\hat{A}$ and $\hat{B}$ is given by:

$$V_{AB}^{\text{cond}} \equiv \langle \hat{A}(0)\hat{B}(0) \rangle_{\text{sym}}^{\text{cond}} - \langle \hat{A}(0) \rangle^{\text{cond}} \langle \hat{B}(0) \rangle^{\text{cond}}$$
$$= \int_0^\infty \frac{d\Omega}{2\pi}\left[\tilde{S}_{AB}(\Omega) - \tilde{G}_A(\Omega)\tilde{G}_B^*(\Omega)\right]. \quad (7.40)$$

Since the first term is the unconditional variance, the second term can be interpreted as reductions in the uncertainty due to acquiring additional information from the measurement.

7.5 Conditional Quantum State and Wiener Filtering

These results can be directly applied to an optomechanical system. Suppose we measure the following quadrature of the cavity output:

$$\hat{Y}_\zeta = \hat{Y}_1 \sin\zeta + \hat{Y}_2 \cos\zeta, \tag{7.41}$$

and its spectral density is:

$$\tilde{S}_{YY}(\Omega) = \tilde{S}_{Y_1 Y_1}(\Omega) \sin^2\zeta + \Re[\tilde{S}_{Y_1 Y_2}(\Omega)] \sin(2\zeta) + \tilde{S}_{Y_2 Y_2}(\Omega) \cos^2\zeta \tag{7.42}$$

The cross-correlation between $\hat{Y}_\zeta$ and the oscillator position (momentum) is simply

$$\tilde{S}_{xY,pY} = \tilde{S}_{xY_1,pY_1}(\Omega) \sin\zeta + \tilde{S}_{xY_2,pY_2}(\Omega) \cos\zeta. \tag{7.43}$$

Substituting for the spectral densities $\tilde{S}_{Y_i Y_j}$, $\tilde{S}_{xY_i,pY_i}$ and $\tilde{S}_{xx,pp}$, derived in Sect. 7.3.2 into Eq. (7.40), we can obtain the conditional covariances of the oscillator position and momentum, namely V_{xx}^{cond}, V_{pp}^{cond} and V_{xp}^{cond}.

To quantify how pure the conditional quantum state is, the occupation number, defined in Eq. (7.32), is no longer an adequate summarizing figure. This is because generally V_{xp}^{cond} is not equal to zero, and a pure squeezed state can have a large occupation number, as defined in Eq. (7.32). A well-defined figure of merit is the *uncertainty product*, which is given by:

$$U \equiv \frac{2}{\hbar}\sqrt{V_{xx}^{\mathrm{cond}} V_{pp}^{\mathrm{cond}} - V_{xp}^{\mathrm{cond}\, 2}}. \tag{7.44}$$

From this, we can introduce an effective occupation number:

$$\mathcal{N}_{\mathrm{eff}} = (U - 1)/2, \tag{7.45}$$

which quantifies how far the quantum state deviates from the pure one with $\mathcal{N}_{\mathrm{eff}} = 0$. This is identical to the previous definition (cf. Eq. (7.32)) in the limiting case of $V_{xx}^{\mathrm{cond}} = V_{pp}^{\mathrm{cond}}/(m^2\omega_m^2)$ and $V_{xp}^{\mathrm{cond}} = 0$, which is actually satisfied in most of the parameter regimes plotted in Fig. 7.1.

For a numerical estimate, and for comparing with the unconditional quantum state in the previous section, we assume the same specification and an ideal phase quadrature detection with $\zeta = 0$ and $\eta = 1$. The resulting effective occupation number is shown in Fig. 7.5. Just as expected, the conditional quantum state is not constrained by the resolved-sideband limit and is almost independent of detailed specifications of γ and Δ. The residual occupation number or impurity of the state, shown in Fig. 7.5, is due to the information of the oscillator motion being confined inside the cavity. Such a confinement is actually attributable to the quantum entanglement between the cavity mode and the oscillator, as we will discuss in Sect. 7.7.

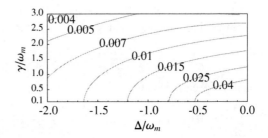

Fig. 7.5 A contour plot for the effective occupation number of the conditional quantum state. For comparison, we have chosen the same specifications as in the unconditional case. From Ref. [37]

7.6 Optimal Feedback Control

Even though the conditional quantum state has minimum variances in position and momentum, the oscillator itself actually wanders around in phase space, with its center given by the conditional mean $[x^{\text{cond}}(t), p^{\text{cond}}(t)]$ at any instant t. In order to localize the mechanical oscillator and achieve its ground state, we need to apply a feedback control, i.e., a force onto the oscillator, according to the measurement results. Such a procedure is shown schematically in Fig. 7.6. Depending on different controllers, the resulting controlled state will have different occupation numbers. The minimum occupation number can only be achieved if the unique optimal controller is applied. In Ref. [11], the optimal controller was derived for a general linear continuous measurement. It can be directly applied to the optomechanical system with non-Markovian quantum noise, as considered here.

Specifically, given the measured output quadrature $\hat{Y}_\zeta$, the feedback force applied to the oscillator can be written in the time and the frequency domains as:

$$\hat{F}_{\text{FB}}(t) = \int_{-\infty}^{t} dt' \, C(t-t') \hat{Y}_\zeta(t'), \text{ and } \tilde{F}_{\text{FB}}(\Omega) = \tilde{C}(\Omega) \tilde{Y}_\zeta(\Omega), \tag{7.46}$$

where $C(t)$ is a causal control kernel. The equation of motion for the oscillator will be modified to (cf. Eq. (7.6)):

$$m[\ddot{\hat{x}}_{\text{ctrl}}(t) + \gamma_m \dot{\hat{x}}_{\text{ctrl}}(t) + \omega_m^2 \hat{x}_{\text{ctrl}}(t)] = -\hbar \bar{G}_0 [\hat{a}^\dagger(t) + \hat{a}(t)] + \hat{\xi}_{\text{th}}(t) + \hat{F}_{\text{FB}}(t). \tag{7.47}$$

In the frequency domain, the controlled oscillator position $\hat{x}_{\text{ctrl}}$ is related to the uncontrolled one $\hat{x}$ by:

$$\tilde{x}_{\text{ctrl}}(\Omega) = \tilde{x}(\Omega) + \frac{\tilde{R}_{xx}^{\text{eff}}(\Omega) \tilde{C}(\Omega) \tilde{Y}_\zeta(\Omega)}{1 - \tilde{R}_{xx}^{\text{eff}}(\Omega) \tilde{R}_{Y_\zeta F}(\Omega) \tilde{C}(\Omega)}. \tag{7.48}$$

As shown in Ref. [11], by minimizing the effective occupation number of the controlled state, the optimal controller can be derived, and it is given by:

$$\tilde{C}^{\text{opt}}(\Omega) = -\frac{\tilde{R}_{xx}^{\text{eff}}(\Omega)^{-1} \tilde{K}_{\text{ctrl}}^{\text{opt}}(\Omega)}{1 - \tilde{R}_{YF}(\Omega) \tilde{K}_{\text{ctrl}}^{\text{opt}}(\Omega)}, \tag{7.49}$$

7.6 Optimal Feedback Control

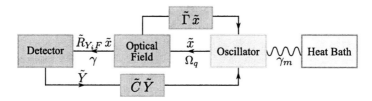

Fig. 7.6 A block diagram for the feedback control scheme. A force is applied onto the mechanical oscillator, based on the measurement result, with a control kernel $\tilde{C}$. In the detuned case ($\Delta \neq 0$,) the radiation pressure and the control force work together to place the mechanical oscillator near its quantum ground state. From Ref. [37]

where

$$\tilde{K}^{\text{opt}}_{\text{ctrl}}(\Omega) = \frac{1}{\tilde{\psi}_+(\Omega)} \left[\tilde{G}_x(\Omega) - \frac{G_x(0)}{\sqrt{V^{\text{cond}}_{pp}/V^{\text{cond}}_{xx}} - i\Omega} \right], \qquad (7.50)$$

with $\tilde{G}_x(\Omega) = [\tilde{S}_{xY}(\Omega)/\tilde{\psi}_-(\Omega)]_+$, as defined in Eq. (7.39).

From Eq. (7.48), we can find out the spectral densities and the covariance for the controlled position and momentum. As it turns out, there is an intimate connection between the optimally-controlled state and the conditional quantum state. Due to the requirement of stationarity, this ensures that $V^{\text{ctrl}}_{xp} = 0$ [$V_{xp} = (1/2)m\dot{V}_{xx}(0) = 0$], and therefore the optimally-controlled state is always less pure than the conditional state. The corresponding purity of the optimally controlled state is [17]

$$U^{\text{opt}}_{\text{ctrl}} = \frac{2}{\hbar} \sqrt{V^{\text{ctrl}}_{xx} V^{\text{ctrl}}_{pp}} \Big|_{\text{optimally controlled}} = \frac{2}{\hbar} \left[\sqrt{V^{\text{cond}}_{xx} V^{\text{cond}}_{pp}} + |V^{\text{cond}}_{xp}| \right]. \qquad (7.51)$$

The occupation number $\mathcal{N}$ for the optimally-controlled state was shown in Fig. 7.1 in the introduction. Since V^{cond}_{xp} is quite small compared with $V^{\text{cond}}_{xx,pp}$, the resulting occupation number is very close to that of the conditional quantum state. Therefore, as long as the optimal controller is applied, the mechanical oscillator is almost in its quantum ground state, and the resolved-sideband limit does not impose significant constraints.

7.7 Conditional Optomechanical Entanglement and Quantum Eraser

In this section, we will analyze the optomechanical entanglement between the oscillator and the cavity mode. In particular, we will show: (i) the residual impurity of the conditional quantum state of the oscillator is induced by this optomechanical entanglement; (ii) if the environmental temperature is high, the existence of entanglement critically depends on whether the information in the cavity output is recovered or not. In other words, the quantum correlation is affected by the "eraser" of certain information, which manifests in the idea of the "quantum eraser" proposed by Scully and Drühl [52].

The existence of optomechanical entanglement is shown in the pioneering work by Vitali et al. [57]. The entanglement criterion, i.e., inseparability, is based upon positivity of the partially transposed density matrix [26, 1, 56]. In the case of Gaussian variables considered here, this reduces to the following uncertainty principle in phase space:

$$\mathbf{V}_{\text{pt}} + \frac{1}{2}\mathbf{K} \geq 0, \quad \mathbf{K} = \begin{pmatrix} 0 & -2i \\ 2i & 0 \end{pmatrix}, \tag{7.52}$$

with $\mathbf{K}$ denoting the commutator matrix. Partial transpose is equivalent to time reversal and the momentum of the oscillator changes sign. The corresponding partially transposed covariance matrix $\mathbf{V}_{\text{pt}} = \mathbf{V}|_{\hat{p} \to -p}$. From the Williamson theorem, there exists a symplectic transformation $S \in S_{p(4,\mathbf{R})}$ such that $\mathbf{S}^T \mathbf{V}_{\text{pt}} \mathbf{S} = \bigoplus_{i=1}^{2} \text{Diag}[\lambda_i, \lambda_i]$. Using the fact that $\mathbf{S}^T \mathbf{K} \mathbf{S} = \mathbf{K}$, the above uncertainty principle requires $\lambda_i \geq 1$. If $\exists \lambda < 1$, the states are entangled. The amount of entanglement can be quantified by the logarithmic negativity $E_\mathcal{N}$ [60, 61], which is defined as

$$E_\mathcal{N} \equiv \max[-\ln \lambda, 0]. \tag{7.53}$$

Given a 4×4 covariance matrix $\mathbf{V}$ between the oscillator $[\hat{x}, \hat{p}]$ and the cavity mode $[\hat{a}_1, \hat{a}_2]$, the simplectic eigenvalue λ has the following closed form:

$$\lambda = \sqrt{\Sigma - \sqrt{\Sigma^2 - 4 \det \mathbf{V}}}/\sqrt{2}, \tag{7.54}$$

where $\Sigma \equiv \det \mathbf{A} + \det \mathbf{B} - 2 \det \mathbf{C}$ and

$$\mathbf{V} = \langle [\hat{x}, \hat{p}, \hat{a}_1, \hat{a}_2]^T [\hat{x}, \hat{p}, \hat{a}_1, \hat{a}_2] \rangle_{\text{sym}} = \begin{bmatrix} \mathbf{A}_{2\times 2} & \mathbf{C}_{2\times 2} \\ \mathbf{C}^T_{2\times 2} & \mathbf{B}_{2\times 2} \end{bmatrix}. \tag{7.55}$$

In Ref. [57], the information contained in the cavity output was ignored and unconditional covariances were used to evaluate the entanglement measure $E_\mathcal{N}$. We can call this unconditional entanglement. If the information were recovered, conditional covariances obtained in Eq. (7.40) would replace the unconditional counterparts. In Fig. 7.7, we compare the unconditional and conditional entanglement. It clearly shows that the entanglement strength increases dramatically in the conditional case. Additionally, the regime where the entanglement is strong is in accord with where the conditional quantum state of the oscillator is less pure as shown in Fig. 7.5. Indeed, there is a simple analytical relation between the effective occupation number $\mathcal{N}_{\text{eff}}$ and the logarithmic negativity $E_\mathcal{N}$ in this ideal case with no thermal noise—that is

$$E_\mathcal{N} = -2 \ln \left[\sqrt{\mathcal{N}_{\text{eff}} + 1} - \sqrt{\mathcal{N}_{\text{eff}}} \right] \approx 2\sqrt{\mathcal{N}_{\text{eff}}}, \tag{7.56}$$

7.7 Conditional Optomechanical Entanglement and Quantum Eraser

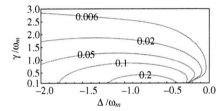
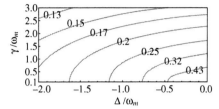

Fig. 7.7 Contour plots of the logarithmic negativity $E_\mathcal{N}$ for unconditional (*left*) and conditional (*right*) entanglement between the cavity mode and the oscillator. We have assumed that $\Omega_q/\omega_m = 0.5$ to make sure that the resulting optomechanical system is stable in the parameter regimes shown in the figure. To manifest the entanglement, we have ignored thermal noise. From Ref. [37]

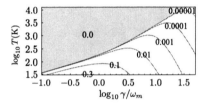

Fig. 7.8 Logarithmic negativity $E_\mathcal{N}$ as a function of cavity bandwidth and environmental temperature. We have chosen $\Omega_q/\omega_m = 1$, $\Delta = 0$, $Q_m = 5 \times 10^5$ and $\omega_m/2\pi = 10^6$ Hz. The shaded regimes are where entanglement vanishes. From Ref. [37]

for small $\mathcal{N}_{\text{eff}}$ [56]. Therefore, the limitation of a cooling experiment actually comes from the optomechanical entanglement, which justifies our claim in Sect. 7.5

If we take into account the environmental temperature as shown in Fig. 7.2 in the introduction part, the unconditional entanglement vanishes when the temperature is higher than 10 K given the following specifications: $\gamma/\omega_m = 1$, $\Delta/\omega_m = -1$, $\Omega_q/\omega_m = 1$ and $Q_m = 5 \times 10^5$ with $\omega_m/2\pi = 10^6$ Hz. In contrast, the conditional one exists even when the temperature becomes higher than 100 K. Therefore, only when the information contained in the cavity output is properly treated will the observer be able to recover the quantum correlation between the oscillator and the cavity mode at high temperature. In fact, the temperature is not the dominant figure that determines the existence of quantum entanglement. A recent investigation showed that, in the simple system with an oscillator interacting with a coherent optical field, quantum entanglement always exists between the oscillator and outgoing optical field [36]. The resulting entanglement strength only depends on the ratio between the characteristic interaction strength Ω_q and the thermal-noise strength Ω_F, rather than on the environmental temperature. We can make some correspondences to the results in Ref. [36] by assuming a large cavity bandwidth. In such a case, the cavity mode exchanges information with the external outgoing field at a timescale much shorter than the thermal decoherence timescale of the oscillator. In Fig. 7.8, we show the resulting $E_\mathcal{N}$ of the conditional entanglement as a function of cavity bandwidth and environmental temperature with fixed interaction strength.

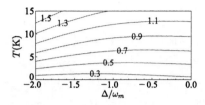

Fig. 7.9 Occupation number of the optimally-controlled state as a function of the temperature and the cavity detuning. The other specifications are chosen to be achievable in a real experiment, and are detailed in the main text. From Ref. [37]

The entanglement can persist at a very high temperature (10^4 K, as shown in this plot!) as long as the cavity bandwidth is large. This, to some extent, recovers the results obtained in Ref. [52].

7.8 Effects of Imperfections and Thermal Noise

To motivate cavity-assisted cooling experiments, we will consider the effects of various imperfections that exist in a real experiment, which include nonunity quantum efficiency of photodetection, thermal noise, and optical loss. The effects of nonunity quantum efficiency and thermal noise have already been taken into account in the equations of motion. With an optical loss, some uncorrelated vacuum fields enter the cavity in an unpredictable way. A small optical loss will not modify the cavity bandwidth significantly but will introduce an additional force noise, which is (cf. Eq. (7.28))

$$S_{FF}^{\text{add}}(\Omega) = 4\hbar m \Omega_q^3 \gamma_\epsilon |\chi|^2 (\gamma^2 + \Omega^2 + \Delta^2), \tag{7.57}$$

where $\gamma_\epsilon \equiv c\epsilon/(4L)$ is the effective bandwidth that is induced by an optical loss of ϵ. For numerical estimations, we will use the following experimentally achievable parameters:

$$m = 1\,\text{mg},\ I_0 = 3\,\text{mW},\ \mathcal{F} = 3 \times 10^4,\ \omega_m/(2\pi) = 10^5\,\text{Hz},$$
$$Q_m = 5 \times 10^6,\ L = 1\,\text{cm},\ \eta = 0.95,\ \epsilon = 10\,\text{ppm}, \tag{7.58}$$

where $\mathcal{F}$ is the cavity finesse, and $Q_m \equiv \omega_m/(\gamma_m)$ is the mechanical quality factor. This gives a coupling strength of $\Omega_q/\omega_m \approx 0.6$ (for $\Delta = -\omega_m$,) and a cavity bandwidth $\gamma/\omega_m = 2.5$. The final results will not change if we increase both the mass and power by the same factor, which essentially gives the same effective interaction strength.

In Fig. 7.9, we show the corresponding occupation number for the controlled state as a function of environmental temperature and cavity detuning. An occupation number less than one can be achieved when the environmental temperature becomes lower than 10 K given the above specifications. If the oscillator can sustain a higher optical power, one can increase the interaction strength to reduce thermal excitations.

7.9 Conclusions

We have shown that both the conditional state and the optimally-controlled state of the mechanical oscillator can achieve a low occupation number, even if the cavity bandwidth is large. Therefore, as long as the information of the oscillator motion contained in the cavity output is carefully recovered, the resolved-sideband limit will not pose a fundamental limit in cavity-assisted cooling experiments. This work can help in the understanding of the intermediate regime between optomechanical cooling and feedback cooling, which will be useful in the search for the optimal parameters for a given experimental setup. In addition, we have shown that the optomechanical entanglement between the cavity mode and the oscillator can be significantly enhanced by recovering information, and its existence becomes insensitive to the environmental temperature.

References

1. G. Adesso, F. Illuminati, Entanglement in continuous-variable systems: recent advances and current perspectives. J. Phys. A: Math. Theor. **40**(28), 7821 (2007)
2. O. Arcizet, P.-F. Cohadon, T. Briant, M. Pinard, A. Heidmann, Radiation-pressure cooling and optomechanical instability of a micromirror. Nature **444**, 71–74 (2006)
3. D.G. Blair, E.N. Ivanov, M.E. Tobar, P.J. Turner, van F. Kann, I.S. Heng, High sensitivity gravitational wave antenna with parametric transducer readout. Phys. Rev. Lett. **74**, 1908 (1995)
4. D. Bouwmeester, A. Ekert, A. Zeilinger, *The Physics of Quantum Information* (Springer, Berlin, 2002)
5. V.B. Braginsky, F.Y. Khalili, *Quantum Measurement* (Cambridge University Press, Cambridge, 1992)
6. A. Buonanno, Y. Chen, Scaling law in signal recycled laserinterferometer gravitational-wave detectors. Phys. Rev. D **67**, 062002 (2003)
7. P.F. Cohadon, A. Heidmann, M. Pinard, Cooling of a mirror by radiation pressure. Phys. Rev. Lett. **83**, 3174 (1999)
8. LIGO Scientific Collaboration, Observation of a kilogram-scale oscillator near its quantum ground state. New J. Phys. **11**(7), 073032 (2009)
9. T. Corbitt, Y. Chen, E. Innerhofer, H. Mueller-Ebhardt, D. Ottaway, H. Rehbein, D. Sigg, S. Whitcomb, C. Wipf, N. Mavalvala, An all-optical trap for a gram-scale mirror. Phys. Rev. Lett. **98**, 150802-4 (2007)
10. T. Corbitt, C. Wipf, T. Bodiya, D. Ottaway, D. Sigg, N. Smith, S. Whitcomb, N. Mavalvala, Optical dilution and feedback cooling of a gram-scale oscillator to 69 mK. Phys. Rev. Lett. **99**, 160801–1608014 (2007)
11. S. Danilishin, H. Mueller-Ebhardt, H. Rehbein, K. Somiya, R. Schnabel, K. Danzmann, T. Corbitt, C. Wipf, N. Mavalvala, Y. Chen, Creation of a quantum scillator by classical control. arXiv:0809.2024 (2008)
12. L. Diosi, A universal master equation for the gravitational violation of quantum mechanics. Phys. Lett. A **120**, 377–381 (1987)
13. L. Diosi, Models for universal reduction of macroscopic quantum fluctuations. Phys. Rev. A **40**, 1165 (1989)
14. L. Diosi, Laser linewidth hazard in optomechanical cooling. Phys. Rev. A **78**, 021801 (2008)
15. A.C. Doherty, K. Jacobs, Feedback control of quantum systems using continuous state estimation. Phys. Rev. A **60**, 2700 (1999)

16. A.C. Doherty, S.M. Tan, A.S. Parkins, D.F. Walls, State determination in continuous measurement. Phys. Rev. A **60**, 2380 (1999)
17. F. Elste, S.M. Girvin, A.A. Clerk, Quantum noise interference and backaction cooling in cavity nanomechanics. Phys. Rev. Lett. **102**, 207209 (2009)
18. I. Favero, C. Metzger, S. Camerer, D. Konig, H. Lorenz, J.P. Kotthaus, K. Karrai, Optical cooling of a micromirror of wavelength size. Appl. Phys. Lett. **90**, 104101–3 (2007)
19. C. Gardiner, P. Zoller, *Quantum Noise* (Springer, Berlin, 2004)
20. C. Genes, D. Vitali, P. Tombesi, Simultaneous cooling and entanglement of mechanical modes of a micromirror in an optical cavity. New J. Phys. **10**(9), 095009 (2008)
21. S. Gigan, H.R. Böhm, M. Paternostro, F. Blaser, G. Langer, J.B. Hertzberg, K.C. Schwab, D. Bäuerle, M. Aspelmeyer, A. Zeilinger, Self-cooling of a micromirror by radiation pressure. Nature **444**, 67–70 (2006)
22. S. Gröblacher, S. Gigan, H.R. Böhm, A. Zeilinger, M. Aspelmeyer, Radiation-pressure self-cooling of a micromirror in a cryogenic environment. EPL Europhys. Lett. **81**(5), 54003 (2008)
23. S. Gröblacher, J.B. Hertzberg, M.R. Vanner, G.D. Cole, S. Gigan, K.C. Schwab, M. Aspelmeyer, Demonstration of an ultracold microoptomechanical oscillator in a cryogenic cavity. Nat. Phys. **5**, 485–488 (2009)
24. M.J. Hartmann, M.B. Plenio, Steady state entanglement in the mechanical vibrations of two dielectric membranes. Phys. Rev. Lett. **101**, 200503 (2008)
25. A. Hopkins, K. Jacobs, S. Habib, K. Schwab, Feedback cooling of a nanomechanical resonator. Phys. Rev. B **68**, 235328 (2003)
26. M. Horodecki, P. Horodecki, R. Horodecki, Separability of mixed states: necessary and sufficient conditions. Phys. Lett. A **223**, 1–8 (1996)
27. G. Jourdan, F. Comin, J. Chevrier, Mechanical mode dependence of bolometric backaction in an atomic force microscopy microlever. Phys. Rev. Lett. **101**, 133904–4 (2008)
28. Y.-H. Kim, R. Yu, S.P. Kulik, Y. Shih, M.O. Scully, Delayed choice quantum eraser. Phys. Rev. Lett. **84**, 1 (2000)
29. H.J. Kimble, Y. Levin, A.B. Matsko, K.S. Thorne, S.P. Vyatchanin, Conversion of conventional gravitational-wave interferometers into quantum nondemolition interferometers by modifying their input and/or output optics. Phys. Rev. D **65**, 022002 (2001)
30. D. Kleckner, D. Bouwmeester, Sub-kelvin optical cooling of a micromechanical resonator. Nature **444**, 75–78 (2006)
31. S. Mancini, D. Vitali, P. Tombesi, Optomechanical cooling of a macroscopic oscillator by homodyne feedback. Phys. Rev. Lett. **80**, 688 (1998)
32. S. Mancini, V. Giovannetti, D. Vitali, P. Tombesi, Entangling macroscopic oscillators exploiting radiation pressure. Phys. Rev. Lett. **88**, 120401 (2002)
33. F. Marquardt, J.P. Chen, A.A. Clerk, S.M. Girvin, Quantum theory of cavity-assisted sideband cooling of mechanical motion. Phys. Rev. Lett. **99**, 093902 (2007)
34. W. Marshall, C. Simon, R. Penrose, D. Bouwmeester, Towards quantum superpositions of a mirror. Phys. Rev. Lett. **91**, 130401 (2003)
35. C.H. Metzger, K. Karrai, Cavity cooling of a microlever. Nature **432**, 1002–1005 (2004)
36. H. Miao, S. Danilishin, Y. Chen, Universal quantum entanglement between an oscillator and continuous fields. arXiv:0908.1053 (2009)
37. H. Miao, S. Danilishin, H. Mller-Ebhardt, Y. Chen, Achieving ground state and enhancing optomechanical entanglement by recovering information. New J. Phys. **12**, 083032 (2010)
38. G.J. Milburn, Classical and quantum conditional statistical dynamics. Quantum Semiclassical Opt.: J. Eur. Opt. Soc. Part B **8**(1), 269 (1996)
39. C.M. Mow-Lowry, A.J. Mullavey, S. Gossler, M.B. Gray, D.E. Mc-Clelland, Cooling of a gram-scale cantilever flexure to 70 mK with a servo-modified optical spring. Phys. Rev. Lett. **100**, 010801–4 (2008)
40. H. Mueller-Ebhardt, H. Rehbein, R. Schnabel, K. Danzmann, Y. Chen, Entanglement of macroscopic test masses and the standard quantum limit in laser interferometry. Phys. Rev. Lett. **100**, 013601 (2008)

References

41. H. Mueller-Ebhardt, H. Rehbein, C. Li, Y. Mino, K. Somiya, R. Schnabel, K. Danzmann, Y. Chen, Quantum-state preparation and macroscopic entanglement in gravitational-wave detectors. Phys. Rev. A **80**, 043802 (2009)
42. A. Naik, O. Buu, M.D. LaHaye, A.D. Armour, A.A. Clerk, M.P. Blencowe, K.C. Schwab, Cooling a nanomechanical resonator with quantum back-action. Nature **443**, 193–196 (2006)
43. A.D. O'Connell, M. Hofheinz, M. Ansmann, R.C. Bialczak, M. Lenander, E. Lucero, M. Neeley, D. Sank, H. Wang, M. Weides, J. Wenner, J.M. Martinis, A.N. Cleland, Quantum ground state and single-phonon control of a mechanical resonator. Nature **464**, 697–703 (2010)
44. M. Paternostro, D. Vitali, S. Gigan, M.S. Kim, C. Brukner, J. Eisert, M. Aspelmeyer, Creating and probing multipartite macroscopic entanglement with light. Phys. Rev. Lett. **99**, 250401 (2007)
45. R. Penrose, On gravity's role in quantum state reduction. Gen. Relativ. Gravitation **28**, 581–600 (1996)
46. A. Peres, Separability criterion for density matrices. Phys. Rev. Lett. **77**, 1413 (1996)
47. M. Poggio, C.L. Degen, H.J. Mamin, D. Rugar, Feedback cooling of a cantilever's fundamental mode below 5 mK. Phys. Rev. Lett. **99**, 017201-4 (2007)
48. P. Rabl, C. Genes, K. Hammerer, M. Aspelmeyer, Phase-noise induced limitations on cooling and coherent evolution in optomechanical systems. Phys. Rev. A **80**, 063819 (2009)
49. S.W. Schediwy, C. Zhao, L. Ju, D.G. Blair, P. Willems, Observation of enhanced optical spring damping in a macroscopic mechanical resonator and application for parametric instability control in advanced gravitationalwave detectors. Phys. Rev. A **77**, 013813-5 (2008)
50. A. Schliesser, P. Del'Haye, N. Nooshi, K.J. Vahala, T.J. Kippenberg, Radiation pressure cooling of a micromechanical oscillator using dynamical backaction. Phys. Rev. Lett. **97**, 243905-4 (2006)
51. A. Schliesser, R. Riviere, G. Anetsberger, O. Arcizet, T.J. Kippenberg, Resolved-sideband cooling of a micromechanical oscillator. Nat. Phys. **4**, 415–419 (2008)
52. M.O. Scully, K. Druhl, Quantum eraser: a proposed photon correlation experiment concerning observation and delayed choice in quantum mechanics. Phys. Rev. A **25**, 2208 (1982)
53. R. Simon, Peres-horodecki separability criterion for continuous variable systems. Phys. Rev. Lett. **84**, 2726 (2000)
54. J.D. Teufel, J.W. Harlow, C.A. Regal, K.W. Lehnert, Dynamical backaction of microwave fields on a nanomechanical oscillator. Phys. Rev. Lett. **101**, 197203-4 (2008)
55. J.D. Thompson, B.M. Zwickl, A.M. Jayich, F. Marquardt, S.M. Girvin, J.G.E. Harris, Strong dispersive coupling of a high-finesse cavity to a micromechanical membrane. Nature **452**, 72–75 (2008)
56. G. Vidal, R.F. Werner, Computable measure of entanglement. Phys. Rev. A **65**, 032314 (2002)
57. D. Vitali, S. Gigan, A. Ferreira, H.R. Böhm, P. Tombesi, A. Guerreiro, V. Vedral, A. Zeilinger, M. Aspelmeyer, Optomechanical entanglement between a movable mirror and a cavity field. Phys. Rev. Lett. **98**, 030405 (2007)
58. S. Vyatchanin, Effective cooling of quantum system. Dokl. Akad. Nauk SSSR **234**, 688 (1977)
59. I. Wilson-Rae, N. Nooshi, W. Zwerger, T.J. Kippenberg, Theory of ground state cooling of a mechanical oscillator using dynamical backaction. Phys. Rev. Lett. **99**, 093901 (2007)
60. C. Zhao, L. Ju, H. Miao, S. Gras, Y. Fan, D.G. Blair, Three-mode optoacoustic parametric amplifier: a tool for macroscopic quantum experiments. Phys. Rev. Lett. **102**, 243902 (2009)
61. W.H. Zurek, Decoherence and the transition from quantum to classical. Phys. Today **44**, 36–44 (1991)

Chapter 8
Universal Entanglement Between an Oscillator and Continuous Fields

8.1 Preface

In the previous two chapters, we have studied the optomechanical entanglement between the optical cavity modes and the mechanical oscillator, both of which have finite degrees of freedom. In this chapter, we study the entanglement between a mechanical oscillator and a coherent continuous optical field which contains infinite degrees of freedom. This system is interesting because it lies in the heart of all optomechanical systems. With a rigorous functional analysis, we develop a new mathematical framework for treating quantum entanglement that involves infinite degrees of freedom. We show that quantum entanglement is always present between the oscillator and the continuous optical field-even when the environmental temperature is high, and the oscillator is highly classical. Such a universal entanglement is also shown to be able to survive more than one mechanical oscillation period, if the characteristic frequency of the optomechanical interaction is larger than that of the thermal noise. In addition, we introduce effective optical modes, which are ordered by their entanglement strength, to better understand the entanglement structure, in analogy with the energy spectrum of an atomic system. In particular, we derive the optical mode that is maximally entangled with the mechanical oscillator, which will be useful for future quantum computing, and for encoding information into mechanical degrees of freedom. This is a joint research effort by Stefan Danilishin, Yanbei Chen and myself. It is published in Phys. Rev. A **81**, 052307 (2010).

8.2 Introduction

Entanglement, as one of the most fascinating features of quantum mechanics, lies in the heart of quantum computing and many quantum communication protocols [2]. Great efforts have been devoted to theoretical and experimental investigations of quantum entanglements in different systems with discrete or continuous variables.

Due to recent significant achievements in fabricating high-Q mechanical oscillators, the quantum entanglement with mechanical degrees of freedom has aroused great interest. Especially, many table-top experiments have demonstrated significant cooling of the mechanical degrees of freedom via feedback or passive damping (self-cooling) [2, 3, 7–10, 14, 19, 20, 21, 24, 25, 29, 32, 34, 38, 40, 41, 42, 45, 46], which in principle allows us to achieve the quantum ground state [11, 12, 18, 26, 28, 39, 49, 51, 52]. More recently, with a conventional cryogenic refrigeration, O'Connell et al. have succeeded in the ground-state cooling of a micromechanical oscillator [35]. These experiments not only illuminate quantum-limited measurements [5], but also pave the way for creating quantum entanglement with mechanical degrees of freedom. Theoretical analysis shows that by coupling a mechanical oscillator to a Fabry-Pérot cavity, one can create stationary (Einstein-Podosky-Rosen) EPR-type quantum entanglement between optical modes and an oscillator [48], or even between two macroscopic oscillators [22, 27]. In Ref. [33], it was shown that entanglement between two oscillators can also be created by conditioning on the continuous measurements of the common and differential optical modes in a laser interferometer.

Here, we consider the quantum entanglement between a mechanical oscillator and a coherent optical field, which models the essential process in all above-mentioned optomechanical systems. There are two important motivations behind this: (1) evaluation of entanglement involving a field which contains infinite degrees of freedom. The entanglement structure itself is an interesting problem. To our knowledge, only finite-degrees-of-freedom entanglements have been investigated in the literature; (2) the effect of thermal decoherence. There is an interesting observation: on the one hand, the environmental temperature enters as an explicit factor and directly determines the existence of the optomechanical entanglement considered in Refs. [20, 48]; on the other hand, only the ratio between the optomechanical interaction and thermal decoherence determines the existence of the entanglement instead of the thermal decoherence alone, and the environmental temperature only influences the entanglement strength implicitly as shown in Refs. [15, 33]. By studying this essential process, we can have a complete picture of the thermal decoherence.

The model and its spacetime diagram are shown schematically in Fig. 8.1. A similar system was analyzed previously by Pirandola et al. [37]. They used a narrow-detection-band approximation to introduce sideband modes, which maps the outgoing field into two effective degrees of freedom. In the situation here, sideband modes are not well-defined, because the interaction turns off at $t = 0$ and only the half-space $[-\infty, 0]$ is involved. Instead, we will directly evaluate the entanglement between the oscillator and the outgoing field $\hat{b}$ (with infinite degrees of freedom), using the positivity of partial transpose (PPT) criterion [1, 13, 23, 36, 43, 44, 47]. Only in the weak-interaction and low-thermal-noise limit can we make correspondences between our results and those obtained in Ref. [37].

The outline of this chapter is as follows: in Sect. 8.3, we will analyze the dynamics of this system, and introduce the covariance between the dynamical quantities, which will be essential for analyzing the quantum entanglement. In Sect. 8.4, we take the continuous limit and extend the PPT criterion to the case with infinite degrees of freedom. In addition, we apply a rigorous functional analysis and obtain the

8.2 Introduction

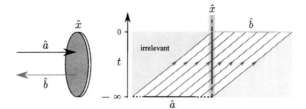

Fig. 8.1 A schematic plot of the model and the corresponding spacetime diagram. Here $\hat{x}$, $\hat{a}$ and $\hat{b}$ denote the oscillator position, ingoing and outgoing field respectively. For clarity, we intentionally place $\hat{a}$ and $\hat{b}$ on different sides of the oscillator world line. The tilted lines represent the light rays. The optical field entering after the moment of interest (t = 0) is out of causal contact, and thus irrelevant. From Ref. [31]

entanglement measure of which a simple scaling is derived. In Sect. 8.5, we study the survival time of the entanglement under thermal decoherence. In Sect. 8.6, we introduce effective optical modes to understand the entanglement structure and obtain the maximally-entangled mode. In Sect. 8.7, we make a numerical estimate given a set of experimental achievable specification to motivate future experiment to investigate entanglement. We conclude with our main results in Sect. 8.8.

8.3 Dynamics and Covariance Matrix

Due to the linearity of the system dynamics, its Heisenberg equations of motion are formally identical to the classical equations of motion, apart from the fact that every dynamical quantity is now treated as a quantum operator. For the optical field, the standard input-output relations are:

$$\hat{b}_1(t) = \hat{a}_1(t), \tag{8.1}$$

$$\hat{b}_2(t) = \hat{a}_2(t) + \kappa \hat{x}(t). \tag{8.2}$$

Here, $\hat{a}_1(\hat{b}_1)$ and $\hat{a}_2(\hat{b}_2)$ are the amplitude and phase quadratures of the ingoing (outgoing) optical field. They are defined through the optical electric field: $\hat{E}_{\text{in}}(t) = \sqrt{\frac{4\pi\hbar\omega_0}{Sc}} \left[(\bar{a} + \hat{a}_1(t)) \cos \omega_0 t + \hat{a}_2(t) \sin \omega_0 t \right]$ which contains a steady-state part $\bar{a}$ and quantum fluctuation parts $\hat{a}_{1,2}$. In this equation, S is the cross-sectional area of the optical beam, $\bar{a} = \sqrt{I_0/(\hbar\omega_0)}$ with I_0 the optical power and ω_0 the laser frequency. A similar relation for $\hat{E}_{\text{out}}$ and $\hat{b}_{1,2}$ also holds. In Eq. (8.2), the displacement of the mechanical oscillator $\hat{x}$ modulates the phase quadrature of the outgoing optical field with an optomechanical coupling constant $\kappa \equiv \omega_0 \bar{a}/c$. To quantify the interaction strength, we introduce a characteristic interaction frequency Ω_q, which is defined by $\Omega_q \equiv \sqrt{\hbar\kappa^2/m}$.

For the mechanical oscillator, the equations of motion are given by:

$$\dot{\hat{x}}(t) = \frac{\hat{p}(t)}{m}, \tag{8.3}$$

$$\dot{\hat{p}}(t) = -\gamma_m \hat{p}(t) - m\omega_m^2 \hat{x}(t) + \hat{F}_{\text{rad}}(t) + \hat{\xi}_{\text{th}}(t). \tag{8.4}$$

Here, $\hat{p}$ is the oscillator momentum. To include the fluctuation-dissipation mechanism of the oscillator coupled to the thermal heat bath at temperature T, we have introduced the mechanical damping γ_m, and the corresponding thermal force noise $\hat{\xi}_{\text{th}}$ which has the following correlation function in the high temperature limit: $\langle \hat{\xi}_{\text{th}}(t) \hat{\xi}_{\text{th}}(t') \rangle = 2m\gamma_m k_B T \delta(t-t') \equiv 2\hbar m \Omega_F^2 \delta(t-t')$ with Ω_F a characteristic frequency of the thermal noise. The presence of thermal noise $\hat{\xi}_{\text{th}}$ ensures the correct commutator between $\hat{x}(t)$ and $\hat{p}(t)$ [16]. The radiation-pressure force $\hat{F}_{\text{rad}}$, up to the first order in the quantum fluctuation, is proportional to $\hat{a}_1(t)$, and $\hat{F}_{\text{rad}}(t) = \hbar\kappa \hat{a}_1(t)$.

The above equations completely quantify the linear dynamics of the system and they can be easily solved. The solution to oscillator position $\hat{x}$ is simply:

$$\hat{x}(t) = \int_{-\infty}^{t} dt' G_x(t-t') \left[\hbar\kappa \hat{a}_1(t') + \hat{\xi}_{\text{th}}(t') \right], \tag{8.5}$$

where the Green's function $G_x(t) \equiv e^{-\gamma_m t} \sin(\omega_m t)/(m\omega_m)$. The radiation-pressure term $\hbar\kappa \hat{a}_1$ induces quantum correlations between the oscillator and the optical field, but it is undermined by $\hat{\xi}_{\text{th}}$. This leads to the question of whether quantum entanglement exists or not, after evolving the entire system from $t = -\infty$ to 0.

Since the variables involved are Gaussian, and linear dynamics will preserve Gaussianity, the quantum entanglement is completely encoded in the covariance matrix $\mathbf{V}$. With the optical field labeled by the continuous time coordinate t, elements of $\mathbf{V}$ involving optical degrees of freedom are defined in the functional space $\mathcal{L}^2[-\infty, 0]$. Specifically,

$$\mathbf{V} = \begin{bmatrix} \mathbf{A} & \mathbf{C}^T \\ \mathbf{C} & \mathbf{B} \end{bmatrix}. \tag{8.6}$$

Here $\mathbf{A}_{ij} = \langle \vec{X}_i \vec{X}_j \rangle_{\text{sym}} (i, j = 1, 2)$, with vector $\vec{X} \equiv [\hat{x}(0), \hat{p}(0)]$ and $\langle \vec{X}_i \vec{X}_j \rangle_{\text{sym}} \equiv \langle \vec{X}_i \vec{X}_j + \vec{X}_j \vec{X}_i \rangle/2$ denoting the symmetrized ensemble average; $\mathbf{C}_{ij}$ and $\mathbf{B}_{ij}$ should be viewed as vectors and operators in $\mathcal{L}^2[-\infty, 0]$. In the coordinate representation, $(t|\mathbf{C}_{ij}) = \langle \vec{X}_i \hat{b}_j(t) \rangle_{\text{sym}}$ and $(t|\mathbf{B}_{ij}|t') = \langle \hat{b}_i(t) \hat{b}_j(t') \rangle_{\text{sym}}$, in which $(|)$ denotes the scalar inner product in $\mathcal{L}^2[-\infty, 0]$.

8.4 Universal Entanglement

According to Refs. [1, 50], in order for one particle and a joint system of arbitrarily large N particles to be separable, a necessary and sufficient condition is that the partially-transposed density matrix $\varrho_{1|N}^{T_1}$ (with respect to the first particle) should be *positive semidefinite*, i.e., $\varrho_{1|N}^{T_1} \geq 0$. In phase space of continuous Gaussian variables, this reduces to the *Uncertainty Principle*:

$$\mathbf{V}_{pt} + \frac{1}{2}\mathbf{K} \geq 0. \tag{8.7}$$

Here, the commutator matrix $\mathbf{K} = \bigoplus_{k=1}^{N+1} 2\sigma_y$ with σ_y denoting a Pauli matrix. According to the Williamson theorem, there exists a symplectic transformation $\mathbf{S} \in S_{p(2N+2,\mathbb{R})}$ such that $\mathbf{S}^T \mathbf{V}_{pt} \mathbf{S} = \bigoplus_{k=1}^{N+1} \text{Diag}[\lambda_k, \lambda_k]$. Using the fact that $\mathbf{S}^T \mathbf{K} \mathbf{S} = \mathbf{K}$, the above *Uncertainty Principle* reads $\lambda_k \geq 1$. If this fails to be the case, i.e., $\exists \lambda_k < 1$, the states are entangled. The amount of entanglement can be quantified by the logarithmic negativity $E_\mathcal{N}$ [47], which is defined as:

$$E_N \equiv \max\left[\sum_k \ln \lambda_k, 0\right] \quad \text{for } k : \lambda_k < 1. \tag{8.8}$$

In the case considered here, N approaches ∞, and the partial transpose is equivalent to time reversal and therefore $\mathbf{V}_{pt} = \mathbf{V}|_{\hat{p}(0) \to -\hat{p}(0)}$. According to Ref. [47], λ_k can be obtained by solving an eigenvalue problem:

$$\mathbf{V}_{pt}\mathbf{v} = \frac{1}{2}\lambda \mathbf{K}\mathbf{v}, \tag{8.9}$$

where $\mathbf{v} \equiv [\alpha_0, \beta_0, |\alpha), |\beta)]^T$ with $|f)$ denoting the vector in $\mathcal{L}^2[-\infty, 0]$. Normalizing $\hat{x}$ and $\hat{p}$ with respect to their zero-point values, the commutator reads $[\hat{x}, \hat{p}] = 2i$. For the optical field, we set $[\hat{b}_1(t), \hat{b}_2(t')] = 2i\,\delta(t - t')$, which gives the coordinate representation of $\mathbf{K}$.

Due to uniqueness of $|\alpha)$ and $|\beta)$ in terms of α_0 and β_0 for any $\lambda < 1$ (non-singular), Eq. (8.9) leads to the following characteristic equation [cf. Eq. (8.6)]

$$\det[\mathbf{A} + \lambda\,\sigma_y - \mathbf{C}^T(\lambda\,\sigma_y + \mathbf{B})^{-1}\mathbf{C}]. = 0 \tag{8.10}$$

It can be shown that:

$$(\lambda\,\sigma_y + \mathbf{B})^{-1} = \begin{bmatrix} 1 + B_\lambda^\dagger M^{-1} B_\lambda & -B_\lambda^\dagger M^{-1} \\ -M^{-1} B_\lambda & M^{-1} \end{bmatrix}, \tag{8.11}$$

where we have used the fact that $B_{12}^\dagger = B_{21}$ in $\mathcal{L}^2[-\infty, 0]$ and have defined $B_\lambda \equiv B_{12} - i\lambda$ and $M \equiv B_{22} - B_\lambda^\dagger B_\lambda$.

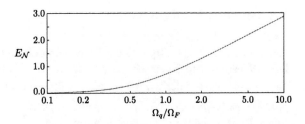

Fig. 8.2 Logarithmic negativity $E_\mathcal{N}$ as a function of the ratio Ω_q/Ω_F. A mechanical quality factor $Q_m = 10^3$ is chosen. From Ref. [31]

To solve this characteristic equation, we need to invert the operator M which can be achieved via the *Wiener-Hopf method*.[1] Given any function $|g\rangle = M^{-1}|h\rangle$, in the frequency domain as:

$$\tilde{g}(\Omega) = \int_{-\infty}^{0} dt\, e^{i\Omega t} M^{-1}|h\rangle = \frac{1}{\tilde{\psi}_-}\left[\frac{\tilde{h}}{\tilde{\psi}_+}\right]_- . \tag{8.12}$$

Here, $[\ ]_-$ means taking the causal part of the given function (i.e., with poles in the lower-half complex plane), and the factorization

$$\tilde{\psi}_+\tilde{\psi}_- \equiv \Lambda + i\lambda\hbar\kappa^2(\tilde{G}_x - \tilde{G}_x^*) + 2\hbar m\kappa^2\Omega_F^2\,\tilde{G}_x\tilde{G}_x^* \tag{8.13}$$

with $\Lambda \equiv 1-\lambda^2$, and $\tilde{G}_x$ denoting the Fourier transformation of $G_x(t)$. In the above equation, $\tilde{\psi}_+(\tilde{\psi}_-)$ and its inverse are analytic in upper-half (lower-half) complex plane, and $\tilde{\psi}_+(-\Omega) = \tilde{\psi}_+^*(\Omega) = \tilde{\psi}_-(\Omega)$. In deriving Eq. (8.13), we have used $\langle \hat{a}_i(t)\hat{a}_j(t')\rangle_{\text{sym}} = \delta_{ij}\delta(t-t')$, and the correlation function for the thermal noise.

Finally, an implicit polynomial equation for the symplectic eigenvalue λ is derived from Eq. (8.10). As it turns out, there always exists one eigenvalue λ whose magnitude is smaller than one. In Fig. 8.2, the corresponding logarithmic negativity [c.f. Eq. (8.8)] is shown as a function of Ω_q/Ω_F. For a high-Q oscillator $Q_m \equiv \omega_m/\gamma_m \gg 1$, up to the leading order of $1/Q_m$, a very elegant expression for $E_\mathcal{N}$ can be derived:

$$E_\mathcal{N} = \frac{1}{2}\ln\left[1 + \frac{25}{8}\frac{\Omega_q^2}{\Omega_F^2}\right]. \tag{8.14}$$

This only depends on the ratio between Ω_q and Ω_F, which clearly indicates the universality of the quantum entanglement. The reason why thermal decoherence (Ω_F) alone determines the existence of entanglement in Refs. [22, 48] originates from the finite transmission of the cavity, and the information of the cavity mode and the oscillator motion leaks into the environment, inducing additional decoherence. This is addressed thoroughly in Ref. [30].

---

[1] An introduction of this method in solving a similar problem can be found in the appendix of Ref. [30].

8.5 Entanglement Survival Duration

We now investigate how long such an entanglement can survive under thermal decoherence. After turning off the optomechanical coupling at $t = 0$, the mechanical oscillator freely evolves for a finite duration τ, driven only by thermal noise. Due to the thermal decoherence, entanglement will gradually vanish. Mathematically, the symplectic eigenvalue will become larger than unity when τ is larger than the survival time τ_s. By replacing $[\hat{x}(0), \hat{p}(0)]$ with $[\hat{x}(\tau), \hat{p}(\tau)]$ and making similar analysis, up to the leading order of $1/Q_m$, we find that τ_s satisfies a transcendental equation:

$$4\Omega_F^4 \theta_s^2 - (2\Omega_F^2 + \Omega_q^2)^2 \sin^2\theta_s - 25\omega_m^4 = 0, \tag{8.15}$$

with $\theta_s \equiv \omega_m \tau_s$. In the case of $\Omega_q < \Omega_F < \omega_m$, the oscillating term can be neglected, leading to:

$$\theta_s = \frac{5}{2}\frac{\omega_m^2}{\Omega_F^2} = \frac{5 Q_m}{2\bar{n}_{\text{th}} + 1}, \tag{8.16}$$

where we have defined the thermal occupation number $\bar{n}_{\text{th}}$ through $k_B T/(\hbar\omega_m) = \bar{n}_{\text{th}} + (1/2)$. Therefore, in this case, if Q_m is larger than $\bar{n}_{\text{th}}$, the entanglement will be able to survive longer than one oscillation period. Since $Q_m > \bar{n}_{\text{th}}$ is also the requirement that the thermal noise induces a momentum diffusion smaller than its zero-point uncertainty [5], this condition is what we intuitively expect. In the strong interaction case with $\Omega_q \gg \Omega_F$, the transcendental equation can be solved numerically, showing that $\theta_s > 1$ is always valid, and that the entanglement can survive at least up to one oscillation period.

8.6 Maximally-Entangled Mode

To gain insight into the structure of this entanglement, we apply the techniques in Ref. [17] and decompose the outgoing field into independent single modes, by convoluting them with some weight functions f_i, namely

$$\hat{O}_i \equiv (f_i|\hat{b}), \ [\hat{O}_i, \hat{O}_j^\dagger] = 2\delta_{ij}, \tag{8.17}$$

which requires $(f_i|f_j) = \delta_{ij}$. If we define $g_{i1} \equiv \Re[f_i]$ and $g_{i2} \equiv \Im[f_i]$, the single-mode quadratures will be

$$\hat{X}_i \equiv (\hat{O}_i + \hat{O}_i^\dagger)/\sqrt{2} = \int_{-\infty}^0 dt\, g_{i1}\,\hat{b}_1 - g_{i2}\,\hat{b}_2, \tag{8.18}$$

$$\hat{Y}_i \equiv (\hat{O}_i - \hat{O}_i^\dagger)/(i\sqrt{2}) = \int_{-\infty}^0 dt\, g_{i2}\,\hat{b}_1 + g_{i1}\,\hat{b}_2. \tag{8.19}$$

Different choices of weight function will generally give optical modes that have different strengths of entanglement with the mechanical oscillator. The function of particular interest is the one that gives an effective optical mode maximally entangled with the oscillator. Using the fact that logarithmic negativity is an entanglement monotone, the optimal weight function can be derived from the following constrained variational equation:

$$\frac{\delta E_{\mathcal{N}}^{\text{sub}}}{\delta g_i} + \mu_i\, g_i = 0 \quad (i = 1, 2), \tag{8.20}$$

where we have neglected unnecessary indices, μ_k is a Lagrange multiplier due to the constraint $(f|f) = 1$, and $E_{\mathcal{N}}^{\text{sub}}$ quantifies the entanglement in the subsystem consisting of the oscillator and the effective optical mode $[\hat{x}(0), \hat{p}(0), \hat{X}, \hat{Y}]$. As it turns out, the optimal weight functions $g_{1,2}$ have the shape of a decay oscillation, with poles ω given by the following polynomial equation:

$$[(\omega - \omega_m)^2 + \gamma_m^2][(\omega + \omega_m)^2 + \gamma_m^2] + \chi = 0, \tag{8.21}$$

where the parameter χ is a functional of $g_{1,2}$, and also depends on Ω_q and Ω_F. Therefore, the weight functions are

$$g_k(t) = A_k\, e^{\gamma_g t} \cos(\omega_g t + \theta_k) \quad (k = 1, 2), \tag{8.22}$$

with γ_g and ω_g being imaginary and real parts of ω. Analytical solutions to parameters A_k, ω_g, γ_g and θ_k require exact expression for χ in terms of g_k, Ω_q and Ω_F, which is rather complicated. Instead, we numerically optimize these parameters to maximize $E_{\mathcal{N}}^{\text{sub}}$.

Taking into account $(f|f) = 1$, A_1 and A_2 can be reduced to a single parameter ζ, which is defined through

$$A_k^2 = \frac{4\gamma_g(\gamma_g^2 + \omega_g^2)\cos^2[\zeta + k(\pi/2)]}{\gamma_g^2 + \omega_g^2 + \gamma_g^2 \cos(2\theta_k) + \gamma_g\omega_g \sin(2\theta_k)}. \tag{8.23}$$

From Eq. (8.21), $\omega_g^2 - \gamma_g^2 = \omega_m^2 - \gamma_m^2$. In addition, a local unitary transformation (rotation and squeezing) will not change the symplectic eigenvalue. Without loss of generality, we can fix $\theta_1 = \pi/2$ and $\theta_2 = 0$. Therefore, only two parameters ω_g and ζ need to be optimized.

In the special case of the weak-interaction and low-thermal-noise limit (Ω_q, $\Omega_F \ll \omega_m$), the optimal ζ_{opt} is equal to $\pi/4$, which indicates $A_1 \approx A_2 = 2\sqrt{\gamma_m}$ for a high-Q oscillator. In addition, as shown in the upper panel of Fig. 8.3, the optimal $\omega_g^{\text{opt}} = \omega_m$, leading to:

$$f(t) = 2\sqrt{\gamma_m}\, e^{\gamma_m t \pm i\, \omega_m t + \phi_0}. \tag{8.24}$$

8.6 Maximally-Entangled Mode

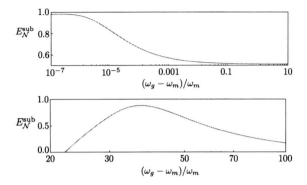

Fig. 8.3 Logarithmic negativity $E_\mathcal{N}^{\text{sub}}$ as a function of quantity $(\omega_g - \omega_m)/\omega_m$ in the weak-interaction and low-thermal-noise case (*upper panel*) and strong-interaction and high-thermal-noise case (*lower panel*). In the first case, we have chosen $Q_m = 10^3$, $\Omega_q/\omega_m = \Omega_F/\omega_m = 2 \times 10^{-2}$. In the second case, $Q_m = 10^6$ (independent of Q_m for higher Q_m), $\Omega_q/\omega_m = 50$, $\Omega_F/\omega_m = 20$ and $\zeta = \pi/3$. From Ref. [31]

Therefore, the optimal weight function has the same shape as the Stokes and Anti-Stokes sideband modes. This is similar to what has been obtained in Refs. [17, 37]; however, due to causality, the weight function here is defined in $\mathcal{L}^2[-\infty, 0]$ rather than in $\mathcal{L}^2[-\infty, \infty]$ which is essential for defining sideband modes.

In the case of strong interaction and high thermal noise ($\Omega_q, \Omega_F > \omega_m$), the optimal ω_g deviates from ω_m and depends on Ω_F and Ω_q, as shown in the lower panel of Fig. 8.3. More generally, the optimal $\zeta_{\text{opt}} = \pi/3$, and ω_g^{opt} can be fitted by $\omega_g^{\text{opt}} \approx (0.64\,\Omega_F^2 + 0.57\,\Omega_q^2)^{1/2}$. Correspondingly, the logarithmic negativity can be approximated as:

$$E_\mathcal{N}^{\text{sub}} \approx \frac{1}{2} \ln\left[1 + \frac{15.\,\Omega_q^2}{13.\,\Omega_F^2 + \Omega_q^2}\right], \tag{8.25}$$

which again manifests universality of the entanglement. Therefore, as long as the optimal weight function is chosen, one can always recover quantum correlations between the oscillator and the outgoing field.

In principle, by choosing a weight function orthogonal to the optimal one obtained above, one can derive the next-order optimal mode. Repeating this procedure will generate a complete spectrum of effective optical modes ordered by $E_\mathcal{N}^{\text{sub}}$, which is analogous to obtaining the wavefunctions and corresponding energy levels with variational method in atomic systems. This not only helps to understand the full entanglement structure but also sheds light on experimental verifications of such universal entanglement. Rather than trying to recover the infinite-dimensional covariance matrix in Eq. (8.6), we can apply right weight functions to extract different effective optical modes and form low-dimension sub-systems. Taking the sub-system consisting of the oscillator and the maximally entangled optical mode for instance, 4×4 covariance matrix can be determined by measuring correlations among different

Table 8.1 Experimental specifications:

	m	$\omega_m/(2\pi)$	Q_m	T	I_0	η
Small scale	50 ng	10^5 Hz	10^7	4 K	0.1 W	0.05
Large scale	40 kg	1 Hz	10^{10}	300 K	800 kW	0.05

quadratures. This can be achieved by using a local oscillator with time-dependent phase, which allows to probe both mechanical quadratures [30] and those of the effective optical mode. For example, a quadrature $\hat{O}_\zeta = \hat{X}\sin\zeta + \hat{Y}\cos\zeta$ can be measured with the following local oscillator light beam:

$$L(t) \propto L_1(t)\cos\omega_0 t + L_2(t)\sin\omega_0 t \tag{8.26}$$

with $L_1(t) = g_1(t)\cos\zeta + g_2(t)\sin\zeta$, and $L_2(t) = g_2(t)\cos\zeta - g_1(t)\sin\zeta$. Synthesis of multiple measurements will recover the covariance matrix that we need to verify the entanglement.

8.7 Numerical Estimates

To motivate future experiments for realizing such a universal entanglement, we will include an important imperfection in a real experiment-the optical loss which comes from the finite transmission of the mirror. This will induce an uncorrelated vacuum field $\hat{n}_{1,2}$ and the input-output relation will be modified to [cf. Eqs. (8.3) and (8.4)]

$$\hat{b}_1(t) = \sqrt{1-\eta}\,\hat{a}_1(t) + \sqrt{\eta}\,\hat{n}_1(t) \tag{8.27}$$

$$\hat{b}_2(t) = \sqrt{1-\eta}\,[\hat{a}_2(t) + \kappa\,\hat{x}(t)] + \sqrt{\eta}\,\hat{n}_2(t) \tag{8.28}$$

with $\eta < 1$ quantifying the optical loss. For a typical optical setup, η can be the order of 0.05 or less. As it turns out, such a small optical loss almost has no effect on the entanglement strength.

To make numerical estimates and demonstrate experimental feasibility, we will consider experimentally achievable specifications for both small-scale and large-scale experiments, which are listed in Table 8.1. For the small scale, the parameters are chosen to be close to that of table-top cooling experiments with micromechanical oscillator, and it gives $\Omega_F/\Omega_q \approx 40$ and $Q_m/\bar{n}_{\text{th}} \approx 10$. For the large scale, it is close to that of an advanced gravitational-wave detector with kg-scale test masses interacting with a high-power optical field [6], and we have $\Omega_F/\Omega_q \approx 1$ and $Q_m/\bar{n}_{\text{th}} \approx 10^{-3}$. In both cases, there is non-vanishing entanglement between the mechanical oscillator and the optical field, and this entanglement can survive up to one mechanical oscillation period.

8.8 Conclusions

We have demonstrated that quantum entanglement exists universally in a system with a mechanical oscillator coupled to continuous optical field. The entanglement measure—logarithmic negativity—displays an elegant scaling which depends on the ratio between characteristic interaction and thermal-noise frequencies. Such scaling should also apply in electromechanical systems, whose dynamics are similar to what we have considered.

References

1. G. Adesso, F. Illuminati, Entanglement in continuous-variable systems: recent advances and current perspectives. J. Phys. A: Math. Theor. **40**(28), 7821 (2007)
2. O. Arcizet, P.-.F. Cohadon, T. Briant, M. Pinard, A. Heidmann, Radiation-pressure cooling and optomechanical instability of a micromirror. Nature **444**, 71–74 (2006)
3. D.G. Blair, E.N. Ivanov, M.E. Tobar, P.J. Turner, van F. Kann, I.S. Heng, High sensitivity gravitational wave antenna with parametric transducer readout. Phys. Rev. Lett. **74**, 1908 (1995)
4. D. Bouwmeester, A. Ekert, A. Zeilinger, *The Physics of Quantum Information* (Springer, Berlin, 2002)
5. V.B. Braginsky, F.Y. Khalili, *Quantum Measurement* (Cambridge University Press, Cambridge, 1992)
6. Z.h.a.o. Chunnong, J.u. Li, M.i.a.o. Haixing, G.r.a.s. Slawomir, F.a.n. Yaohui, B.l.a.i.r. David G., Three-mode optoacoustic parametric amplifier: a tool for macroscopic quantum experiments. Phys. Rev. Lett. **102**, 243902 (2009)
7. P.F. Cohadon, A. Heidmann, M. Pinard, Cooling of a mirror by radiation pressure. Phys. Rev. Lett. **83**, 3174 (1999)
8. LIGO Scientific Collaboration, Observation of a kilogram-scale oscillator near its quantum ground state. New J. Phys. **11**(7), 073032 (2009)
9. T. Corbitt, Y. Chen, E. Innerhofer, H. Mueller-Ebhardt, D. Ottaway, H. Rehbein, D. Sigg, S. Whitcomb, C. Wipf, N. Mavalvala, An all-optical trap for a gram-scale mirror. Phys. Rev. Lett. **98**, 150802–4 (2007)
10. T. Corbitt, C. Wipf, T. Bodiya, D. Ottaway, D. Sigg, N. Smith, S. Whitcomb, N. Mavalvala, Optical dilution and feedback cooling of a gram-scale oscillator to 6.9 mK. Phys. Rev. Lett. **99**, 160801–4 (2007)
11. S. Danilishin, H. Mueller-Ebhardt, H. Rehbein, K. Somiya, R. Schnabel, K. Danzmann, T. Corbitt, C. Wipf, N. Mavalvala, Y. Chen, Creation of a quantum oscillator by classical control. arXiv:0809.2024, 2008
12. L. Diosi, Laser linewidth hazard in optomechanical cooling. Phys. Rev. A **78**, 021801 (2008)
13. L.-.M. Duan, G. Giedke, J.I. Cirac, P. Zoller, Inseparability criterion for continuous variable systems. Phys. Rev. Lett. **84**, 2722 (2000)
14. I. Favero, C. Metzger, S. Camerer, D. Konig, H. Lorenz, J.P. Kotthaus, K. Karrai, Optical cooling of a micromirror of wavelength size. Appl. Phys. Lett. **90**, 104101–3 (2007)
15. A. Ferreira, A. Guerreiro, V. Vedral, Macroscopic thermal entanglement due to radiation pressure. Phys. Rev. Lett. **96**, 060407 (2006)
16. C. Gardiner, P. Zoller, *Quantum Noise* (Springer, Berlin, 2004)
17. C. Genes, A. Mari, P. Tombesi, D. Vitali, Robust entanglement of a micromechanical resonator with output optical fields. Phys. Rev. A **78**, 032316 (2008)
18. C. Genes, D. Vitali, P. Tombesi, Simultaneous cooling and entanglement of mechanical modes of a micromirror in an optical cavity. New J. Phys. **10**(9), 095009 (2008)

19. S. Gigan, H.R. Böhm, M. Paternostro, F. Blaser, G. Langer, J.B. Hertzberg, K.C. Schwab, D. Bäuerle, M. Aspelmeyer, A. Zeilinger, Self-cooling of a micromirror by radiation pressure. Nature **444**, 67–70 (2006)
20. S. Gröblacher, S. Gigan, H.R. Böhm, A. Zeilinger, M. Aspelmeyer, Radiation-pressure self-cooling of a micromirror in a cryogenic environment. EPL (Europhys. Lett.) **81**(5), 54003 (2008)
21. S. Gröblacher, J.B. Hertzberg, M.R. Vanner, G.D. Cole, S. Gigan, K.C. Schwab, M. Aspelmeyer, Demonstration of an ultracold microoptomechanical oscillator in a cryogenic cavity. Nat. Phys. **5**, 485–488 (2009)
22. M.J. Hartmann, M.B. Plenio, Steady state entanglement in the mechanical vibrations of two dielectric membranes. Phys. Rev. Lett. **101**, 200503 (2008)
23. M. Horodecki, P. Horodecki, R. Horodecki, Separability of mixed states: necessary and sufficient conditions. Phys. Lett. A **223**, 1–8 (1996)
24. G. Jourdan, F. Comin, J. Chevrier, Mechanical mode dependence of bolometric backaction in an atomic force microscopy microlever. Phys. Rev. Lett. **101**, 133904–4 (2008)
25. D. Kleckner, D. Bouwmeester, Sub-kelvin optical cooling of a micromechanical resonator. Nature **444**, 75–78 (2006)
26. S. Mancini, D. Vitali, P. Tombesi, Optomechanical cooling of a macroscopic oscillator by homodyne feedback. Phys. Rev. Lett. **80**, 688 (1998)
27. S. Mancini, V. Giovannetti, D. Vitali, P. Tombesi, Entangling macroscopic oscillators exploiting radiation pressure. Phys. Rev. Lett **88**, 120401 (2002)
28. F. Marquardt, J.P. Chen, A.A. Clerk, S.M. Girvin, Quantum theory of cavity-assisted sideband cooling of mechanical motion. Phys. Rev. Lett. **99**, 093902 (2007)
29. C.H. Metzger, K. Karrai, Cavity cooling of a microlever. Nature **432**, 1002–1005 (2004)
30. H. Miao, S. Danilishin, H. Mueller-Ebhardt, H. Rehbein, K. Somiya, Y. Chen, Probing macroscopic quantum states with a sub-Heisenberg accuracy. Phys. Rev. A **81**, 012114 (2010)
31. H. Miao, S. Danilishin, Y. Chen Universal quantum entanglement between an oscillator and continuous fields. arXiv:0908.1053, 2009
32. C.M. Mow-Lowry, A.J. Mullavey, S. Gossler, M.B. Gray, D.E. Mc- Clelland, Cooling of a gram-scale cantilever flexure to 70 mK with a servo-modified optical spring. Phys. Rev. Lett. **100**, 010801–4 (2008)
33. H. Mueller-Ebhardt, H. Rehbein, R. Schnabel, K. Danzmann, Y. Chen, Entanglement of macroscopic test masses and the standard quantum limit in laser interferometry. Phys. Rev. Lett. **100**, 013601 (2008)
34. A. Naik, O. Buu, M.D. LaHaye, A.D. Armour, A.A. Clerk, M.P. Blencowe, K.C. Schwab, Cooling a nanomechanical resonator with quantum back-action. Nature **443**, 193–196 (2006)
35. A.D. O'Connell, M. Hofheinz, M. Ansmann, R.C. Bialczak, M. Lenander, E. Lucero, M. Neeley, D. Sank, H. Wang, M. Weides, J. Wenner, J.M. Martinis, A.N. Cleland, Quantum ground state and single-phonon control of a mechanical resonator. Nature **464**, 697–703 (2010)
36. A. Peres, Separability criterion for density matrices. Phys. Rev. Lett. **77**, 1413 (1996)
37. S. Pirandola, S. Mancini, D. Vitali, P. Tombesi, Continuous-variable entanglement and quantum-state teleportation between optical and macroscopic vibrational modes through radiation pressure. Phys. Rev. A **68**, 062317 (2003)
38. M. Poggio, C.L. Degen, H.J. Mamin, D. Rugar, Feedback cooling of a cantilever's fundamental mode below 5 mK. Phys. Rev. Lett **99**, 017201–4 (2007)
39. P. Rabl, C. Genes, K. Hammerer, M. Aspelmeyer, Phase-noise induced limitations on cooling and coherent evolution in optomechanical systems. Phys. Rev. A **80**, 063819 (2009)
40. S.W. Schediwy, C. Zhao, L. Ju, D.G. Blair, P. Willems, Observation of enhanced optical spring damping in a macroscopic mechanical resonator and application for parametric instability control in advanced gravitationalwave detectors. Phys. Rev. A **77**, 013813–5 (2008)
41. A. Schliesser, P. Del'Haye, N. Nooshi, K.J. Vahala, T.J. Kippenberg, Radiation pressure cooling of a micromechanical oscillator using dynamical backaction. Phys. Rev. Lett. **97**, 243905–4 (2006)

References

42. A. Schliesser, R. Riviere, G. Anetsberger, O. Arcizet, T.J. Kippenberg, Resolved-sideband cooling of a micromechanical oscillator. Nat. Phys. **4**, 415–419 (2008)
43. A. Serafini, Multimode uncertainty relations and separability of continuous variable states. Phys. Rev. Lett. **96**, 110402 (2006)
44. R. Simon, Peres-horodecki separability criterion for continuous variable systems. Phys. Rev. Lett. **84**, 2726 (2000)
45. J.D. Teufel, J.W. Harlow, C.A. Regal, K.W. Lehnert, Dynamical backaction of microwave fields on a nanomechanical oscillator. Phys. Rev. Lett. **101**, 197203–4 (2008)
46. J.D. Thompson, B.M. Zwickl, A.M. Jayich, F. Marquardt, S.M. Girvin, J.G.E. Harris, Strong dispersive coupling of a high-finesse cavity to a micromechanical membrane. Nature **452**, 72–75 (2008)
47. G. Vidal, R.F. Werner, Computable measure of entanglement. Phys. Rev. A **65**, 032314 (2002)
48. D. Vitali, S. Gigan, A. Ferreira, H.R. Böhm, P. Tombesi, A. Guerreiro, V. Vedral, A. Zeilinger, M. Aspelmeyer, Optomechanical entanglement between a movable mirror and a cavity field. Phys. Rev. Lett. **98**, 030405 (2007)
49. S. Vyatchanin, Effective cooling of quantum system, Dokl. Akad. Nauk SSSR **234**, 688 (1977)
50. R.F. Werner, M.M. Wolf, Bound entangled gaussian states. Phys. Rev. Lett. **86**, 3658 (2001)
51. I. Wilson-Rae, N. Nooshi, W. Zwerger, T.J. Kippenberg, Theory of ground state cooling of a mechanical oscillator using dynamical backaction. Phys. Rev. Lett. **99**, 093901 (2007)
52. C. Zhao, L. Ju, H. Miao, S. Gras, Y. Fan, D.G. Blair, Three-mode optoacoustic parametric amplifier: a tool for macroscopic quantum experiments. Phys. Rev. Lett. **102**, 243902 (2009)

Chapter 9
Nonlinear Optomechanical System for Probing Mechanical Energy Quantization

9.1 Preface

In the previous chapters, we have been discussing linear optomechanical devices, of which the dynamics are quantified by linear equations of motion. Motivated by the pioneering theoretical work of Santamore [14], Martin and Zureck [10], and by recent experimental work of Thompson et al. [16], we consider the quantum limit for probing mechanical energy quantization with mechanical modes parametrically coupled to external degrees of freedom. We find that the resolution of a single mechanical quantum requires a strong-coupling regime—the decay rate of external degrees of freedom should be smaller than the parametric coupling rate. In the case of cavity-assisted optomechanical systems, the zero-point motion of the mechanical oscillator needs to be comparable to the linear dynamical range of the optical system, which is characterized by the optical wavelength divided by the cavity finesse. If this condition is satisfied, the nonlinearity of the optomechanical system will become important in the quantum regime. Since a direct probe of the mechanical energy naturally means that we are able to create a non-Gaussian Fock state in the mechanical oscillator, this condition also sets the requirement for creating a non-Gaussian state in optomechanical devices. This is a collaborative research effort by Stefan Danilishin, Thomas Corbitt, Yanbei Chen, and myself. It is published in Phys. Rev. Lett. **103**, 1000402 (2009).

9.2 Introduction

Recently, significant cooling of mechanical modes of harmonic oscillators has been achieved by extracting heat through parametric damping or active feedback [13, 16]. Theoretical calculations suggest that oscillators with a large thermal occupation number ($k_B T \gg \hbar \omega_m$) can be cooled to be close to their ground state, if they have high enough quality factors [11]. Once the ground state is approached, many interesting studies of macroscopic quantum mechanics can be performed, e.g. teleporting a quantum state onto mechanical degrees of freedom [9], creating quantum entanglement between a cavity mode and an oscillator [8] and between two macroscopic test masses [17]. Most proposals involve the oscillator position being linearly coupled to photons, in which case the quantum features of the oscillator, to a great extent, are attributable to the quantization of photons. In order to probe the intrinsic quantum nature of an oscillator, one of the most transparent approaches is to directly measure its energy quantization, and the quantum jumps between discrete energy eigenstates. Since linear couplings alone will not project an oscillator onto its energy eigenstates, nonlinearities are generally required [10, 12, 14]. For cavity-assisted optomechanical systems, one experimental scheme, proposed in the pioneering work of Thompson et al. [16], is to place a dielectric membrane inside a high-finesse Fabry-Pérot cavity, forming a pair of coupled cavities.[1] If the membrane is appropriately located, a dispersive coupling between the membrane position and the optical field is predominantly quadratic, allowing the detection of mechanical energy quantization.

In this chapter, we show that in the experimental setup of Thompson et al. the optical field also couples linearly to the membrane. Due to the finiteness of cavity finesse (either intentional for readout or due to optical losses), this linear coupling introduces quantum back-action. Interestingly, it sets forth a simple Standard Quantum Limit (SQL), which dictates that only those systems whose cavity-mode decay rates are smaller than the optomechanical coupling rate can successfully resolve energy levels. We will further show that a similar constraint applies universally to all experiments that attempt to probe mechanical energy quantization via parametric coupling with external degrees of freedom (either optical or electrical).

9.3 Coupled Cavities

The optical configuration of the coupled cavities is shown in Fig. 9.1. Given the specifications in Ref. [16], the transmissivities of the membrane and end mirrors are quite low, and thus a two-mode description is appropriate [1, 7], with the corresponding Hamiltonian:

$$\hat{\mathcal{H}} = \hbar \omega_m (\hat{q}^2 + \hat{p}^2)/2 + \hbar \omega_0 (\hat{a}^\dagger \hat{a} + \hat{b}^\dagger \hat{b}) - \hbar \omega_s (\hat{a}^\dagger \hat{b} + \hat{b}^\dagger \hat{a}) \\ + \hbar G_0 \hat{q} (\hat{a}^\dagger \hat{a} - \hat{b}^\dagger \hat{b}) + \hat{\mathcal{H}}_{\text{ext}} + \hat{\mathcal{H}}_\xi. \tag{9.1}$$

Here, $\hat{q}, \hat{p}$ are the normalized position and momentum of the membrane; $\hat{a}, \hat{b}$ are annihilation operators of cavity modes in the individual cavities (both resonate at ω_0); $\omega_s \equiv t_m c/L$ is the optical coupling constant for $\hat{a}$ and $\hat{b}$, through the transmission of the membrane [7]; $G_0 \equiv 2\sqrt{2}\omega_0 x_q/L$ is the optomechanical

---

[1] A similar configuration has been proposed by Braginsky et al. for detecting gravitational-waves, Phys. Lett. A **232**, 340 (1997) and Phys. Lett. A **246**, 485 (1998).

9.3 Coupled Cavities

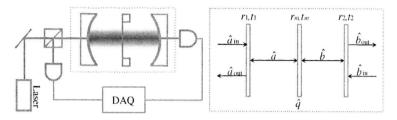

Fig. 9.1 The *left panel* presents the schematic configuration of coupled cavities in the proposed experiment [3]. The *right panel* shows the optical modes, and we denote the reflectivity and transmissivity of the optical elements by r_i and t_i ($i = 1, 2, m$). From Ref. [10]

coupling constant with L denoting the cavity length, and the zero-point motion is $x_q \equiv \sqrt{\hbar/(2m\omega_m)}$; $\hat{\mathcal{H}}_{\text{ext}}$ and $\hat{\mathcal{H}}_\xi$ correspond to the coupling of the system to the environment and quantify the fluctuation and dissipation mechanism. By introducing *optical* normal modes, namely the common mode $\hat{c} \equiv (\hat{a} + \hat{b})/\sqrt{2}$ and differential mode $\hat{d} \equiv (\hat{a} - \hat{b})/\sqrt{2}$,

$$\hat{\mathcal{H}}/\hbar = \frac{\omega_m}{2}(\hat{q}^2 + \hat{p}^2) + \omega_- \hat{c}^\dagger \hat{c} + \omega_+ \hat{d}^\dagger \hat{d} + G_0 \hat{q}(\hat{c}^\dagger \hat{d} + \hat{d}^\dagger \hat{c})$$
$$+ i(\sqrt{2\gamma_c}\, \hat{c}^\dagger \hat{c}_{\text{in}} + \sqrt{2\gamma_d}\, \hat{d}^\dagger \hat{d}_{\text{in}} - \text{H.c.}) + \hat{\mathcal{H}}_\xi/\hbar \quad (9.2)$$

where $\omega_\pm \equiv \omega_0 \pm \omega_s$ and in the Markovian approximation $\hat{\mathcal{H}}_{\text{ext}}$ is written out explicitly in the second line (with $\gamma_{c,d}$ denoting decay rates, and H.c. for Hermitian conjugate).

Before analyzing the detailed dynamics, here we follow Thompson et al. [16] and Bhattacharya and Meystre [1], by assuming $\omega_m \ll \omega_s$ and $G_0 \ll |\omega_+ - \omega_-| = 2\omega_s$, analogous to the dispersive regime in a photon-number counting experiment with a superconducting qubit [2, 15]. This allows us to treat $\hbar G_0 \hat{q}(\hat{c}^\dagger \hat{d} + \hat{d}^\dagger \hat{c})$ as a perturbation, and diagonalize the Hamiltonian formally. Up to $G_0^2/(2\omega_s)^2$, the optical and optomechanical coupling parts of the original Hamiltonian can be written as:

$$\hat{\mathcal{H}}/\hbar = \left(\omega_- - \frac{G_0^2 \hat{q}^2}{2\omega_s}\right)\hat{o}^\dagger \hat{o} + \left(\omega_+ + \frac{G_0^2 \hat{q}^2}{2\omega_s}\right)\hat{e}^\dagger \hat{e}. \quad (9.3)$$

At first sight, the frequency shift of the eigenmodes $\hat{o}$ and $\hat{e}$ is proportional to $\hat{q}^2$. Since the frequency separation of the two normal modes is $2\omega_s \gg \gamma_{c,d}$, they can be independently driven and detected. Besides, with $\gamma_{c,d} < \omega_m$, only the averaged membrane motion is registered and, $\overline{\hat{q}^2} = \hat{N} + 1/2$, with $\hat{N}$ denoting the number of quanta. Therefore, previous authors had concluded that such a *purely* dispersive coupling allows quantum non-demolition (QND) measurements of the mechanical quanta.

However, the new eigenmodes $\hat{o}$ and $\hat{e}$ are given by:

$$\hat{o} = \hat{c} - [(G_0 \hat{d})/(2\omega_s)]\hat{q}, \quad \hat{e} = \hat{d} + [(G_0 \hat{c})/(2\omega_s)]\hat{q}. \quad (9.4)$$

If we pump $\hat{c}$ with classical amplitude $\bar{c}$, and left $\hat{d}$ in vacuum state, the detected mode $\hat{o}$ will have a negligible linear response. However, the idle mode $\hat{e} \approx [G_0 \bar{c}/(2\omega_s)]\hat{q}$, which is dominated by linear coupling. If we choose to drive $\hat{d}$, the role of $\hat{o}$ and $\hat{e}$ will simply interchange. Such linear coupling can potentially demolish the energy eigenstates that we wish to probe. We can make an order-of-magnitude estimate. The optomechanical coupling term in Eq. (9.2), at the linear order, is $G_0 \hat{q}(\bar{c}\hat{d} + \bar{c}^*\hat{d}^\dagger)$. According to the Fermi's golden rule, it causes decoherence of energy eigenstate near the ground level at a rate of:

$$\tau_{\text{dec}}^{-1} = G_0^2 |\bar{c}|^2 \tilde{S}_{\hat{d}}(-\omega_m) \approx G_0^2 |\bar{c}|^2 \gamma_d/(2\omega_s^2), \tag{9.5}$$

where we have assumed that $\hat{c}$ is on resonance, and

$$\tilde{S}_{\hat{d}} \equiv \int dt\, e^{i\omega t} \langle \hat{d}(t)\hat{d}^\dagger(0)\rangle = 2\gamma_d/[(\omega - 2\omega_s)^2 + \gamma_d^2]. \tag{9.6}$$

On the other hand, from Eq. (9.3) and linear response theory [3], the measurement time scale to resolve the energy eigenstate (i.e. measuring $\hat{N}$ with a unit error) with a shot-noise limited sensitivity is approximately given by:

$$\tau_m \approx [\gamma_c^2 \omega_s^2/(G_0^4|\bar{c}|^2)]\tilde{S}_{\hat{c}}(0) = 2\omega_s^2 \gamma_c/(G_0^4|\bar{c}|^2), \tag{9.7}$$

where $\tilde{S}_{\hat{c}}(0)$ is the spectral density of $\hat{c}$ at zero frequency. Requiring $\tau_m \le \tau_{\text{dec}}$ yields

$$(\gamma_c \gamma_d / G_0^2) \lesssim 1. \tag{9.8}$$

In the case when the transmissivity of the end mirrors $t_1 = t_2 \equiv t_0$, we have $\gamma_c = \gamma_d = c\, t_0^2/(2L)$. Defining the cavity finesse as $\mathcal{F} \equiv \pi/t_0^2$, the above inequality reduces to $\lambda/(\mathcal{F} x_q) \lesssim 8\sqrt{2}$. Therefore, to probe mechanical energy quantization, it requires a strong-coupling regime (cf. Eq. (9.8)), or equivalently, for such an optomechanical system, zero-point mechanical motion x_q to be comparable to the linear dynamical range $\lambda/\mathcal{F}$ of the cavity.

We now carry out a detailed analysis of the dynamics according to the standard input-output formalism [5]. In the rotating frame at the laser frequency ω_+, the nonlinear quantum Langevin equations are given by

$$\dot{\hat{q}} = \omega_m \hat{p}, \tag{9.9}$$

$$\dot{\hat{p}} = -\omega_m \hat{q} - \gamma_m \hat{p} - G_0(\hat{c}^\dagger \hat{d} + \hat{d}^\dagger \hat{c}) + \xi_{\text{th}}, \tag{9.10}$$

$$\dot{\hat{c}} = -\gamma_c \hat{c} - i G_0 \hat{q}\, \hat{d} + \sqrt{2\gamma_c}\, \hat{c}_{\text{in}}, \tag{9.11}$$

$$\dot{\hat{d}} = -(\gamma_d + 2i\omega_s) \hat{d} - i G_0 \hat{q}\, \hat{c} + \sqrt{2\gamma_d}\, \hat{d}_{\text{in}}. \tag{9.12}$$

Here, the mechanical damping and associated Brownian thermal force ξ_{th} originate from $\hat{\mathcal{H}}_\xi$ under the Markovian approximation. These equations can be solved

9.3 Coupled Cavities

perturbatively by decomposing every Heisenberg operator $\hat{\alpha}$ into different orders, such that $\hat{\alpha} = \bar{\alpha} + \epsilon \hat{\alpha}^{(1)} + \epsilon^2 \hat{\alpha}^{(2)} + \mathcal{O}[\epsilon^3]$. We treat $G_0/(2\omega_s)$, and the vacuum fluctuations $\sqrt{2\gamma_c}\,\hat{c}_{\text{in}}^{(1)}$ and $\sqrt{2\gamma_d}\,\hat{d}_{\text{in}}^{(1)}$ (simply denoted by $\sqrt{2\gamma_c}\,\hat{c}_{\text{in}}$ and $\sqrt{2\gamma_d}\,\hat{d}_{\text{in}}$ in later discussions) as being of the order of ϵ ($\epsilon \ll 1$).

To the zeroth order, $\bar{c} = \sqrt{2I_0/(\gamma_c \hbar \omega_0)}$, with I_0 denoting the input optical power, and $\bar{d} = 0$. Up to the first order, the radiation pressure term reads $G_0 \bar{c}[\hat{d}^{(1)} + \hat{d}^{(1)\dagger}]$ ($\bar{c}$ is set to be real by choosing an appropriate phase reference). In the frequency domain, it can be written as:

$$\tilde{F}_{\text{rp}} = \frac{2\sqrt{\gamma_d}\,G_0\,\bar{c}[(\gamma_d - i\omega)\tilde{v}_1 - 2\omega_s \tilde{v}_2] + 4G_0^2 \bar{c}^2 \omega_s \tilde{q}}{(\omega + 2\omega_s + i\gamma_d)(\omega - 2\omega_s + i\gamma_d)}, \qquad (9.13)$$

where $\tilde{v}_1$, $\tilde{v}_2$ and $\tilde{q}$ are the Fourier transformations of $\hat{v}_1(t) \equiv (\hat{d}_{\text{in}} + \hat{d}_{\text{in}}^{\dagger})/\sqrt{2}$, $\hat{v}_2(t) \equiv (\hat{d}_{\text{in}} - \hat{d}_{\text{in}}^{\dagger})/(i\sqrt{2})$ and $\hat{q}(t)$, respectively. The first part, containing vacuum fluctuations, is the back-action $\hat{F}_{\text{BA}}$, which induces the quantum limit. The other part proportional to $\tilde{q}$ is due to the optical-spring effect. Within the time scale for measuring energy quantization, of the order of γ_c^{-1} ($\ll \gamma_m^{-1}$), the positive damping can be neglected but the negative rigidity has an interesting consequence—it modifies ω_m to an effective ω_{eff} ($<\omega_m$). Correspondingly, the position of the high-Q membrane is

$$\hat{q}(t) = \hat{q}_m + \Lambda^2 \int_0^t dt'\, \sin\omega_{\text{eff}}(t - t')[\hat{F}_{\text{BA}}(t') + \xi_{\text{th}}(t')], \qquad (9.14)$$

with $\Lambda \equiv \sqrt{\omega_m/\omega_{\text{eff}}}$. The free quantum oscillation $\hat{q}_m = \Lambda(\hat{q}_0 \cos\omega_{\text{eff}} t + \hat{p}_0 \sin\omega_{\text{eff}} t)$ and $\hat{q}_0$ and $\hat{p}_0$ are the initial position and momentum normalized with respect to $\sqrt{\hbar/(m\omega_{\text{eff}})}$ and $\sqrt{\hbar m \omega_{\text{eff}}}$, respectively.

The dispersive response is given by the second-order perturbation $\mathcal{O}[\epsilon^2]$. Adiabatically eliminating rapidly oscillating components and assuming $\omega_m \ll \omega_s$ which can be shown to maximize the signal-to-noise ratio, we obtain

$$\hat{c}^{(2)}(t) = -iG_0 \int_0^t dt'\, e^{-\gamma_c(t-t')} \hat{q}(t') \hat{d}^{(1)}(t')$$
$$\approx G_{\text{eff}}^2\,\bar{c}\,\hat{N}(t)/(2i\gamma_c \omega_s). \qquad (9.15)$$

Here $G_{\text{eff}} \equiv \Lambda G_0$ and $\hat{N}(t) \equiv \hat{N}_0 + \Delta \hat{N}(t)$ contains the number of mechanical quanta $\hat{N}_0 \equiv (\hat{q}_0^2 + \hat{p}_0^2)/2$ and the noise term $\Delta \hat{N}(t)$ due to the back-action and thermal noise. To read out $\hat{N}(t)$, we integrate output phase quadrature for a duration τ. According to the input-output relation $\hat{c}_{\text{out}} + \hat{c}_{\text{in}} = \sqrt{2\gamma_c}\,\hat{c}$, the estimator reads:

$$\hat{Y}(\tau) = \int_0^\tau dt[\hat{u}_2(t) - G_{\text{eff}}^2\,\bar{c}\,\hat{N}(t)/(\sqrt{\gamma_c}\,\omega_s)], \qquad (9.16)$$

where $\hat{u}_2 \equiv (\hat{c}_{\text{in}} - \hat{c}_{\text{in}}^{\dagger})/(i\sqrt{2})$. For Gaussian and Markovian process, the correlation function $\langle \hat{c}_2(t) \hat{c}_2^{\dagger}(t') \rangle = \delta(t - t')/2$. For typical experiments, the thermal occupation number $\bar{n}_{\text{th}} \equiv k_B T/(\hbar \omega_m)$ is much larger than unity, and $\langle \xi_{\text{th}}(t) \xi_{\text{th}}(t') \rangle \approx$

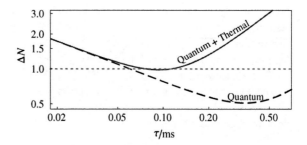

Fig. 9.2 The resolution ΔN for measuring mechanical energy quantization depending on the integration duration τ with total noise (Solid) and quantum noise only (Dashed). From Ref. [6]

$2\gamma_m \bar{n}_{\text{th}} \delta(t - t')$. Through evaluating the four-point correlation function of back-action noise and $\xi_{\text{th}}(t)$ in $\langle \Delta \hat{N}(t) \Delta \hat{N}(t') \rangle$, we obtain the resolution ΔN as a function of τ:

$$\Delta N^2 = \left(\frac{\gamma_c \omega_s^2}{G_{\text{eff}}^4 \bar{c}^2 \tau}\right) + \frac{5}{6}\left(\frac{\gamma_d G_{\text{eff}}^2 \bar{c}^2 \tau}{2\sqrt{2}\,\omega_s^2}\right)^2 + \frac{5}{6}\left(\frac{\gamma_m k_B T \tau}{\sqrt{2}\,\hbar \omega_{\text{eff}}}\right)^2. \quad (9.17)$$

In order to successfully observe energy quantization, the following conditions are simultaneously required: (i) the resolution ΔN^2 should have a minimum equal or less than unity. (ii) this minimum should be reachable within τ that is longer than the cavity storage time $1/\gamma_c$ (which in turn must be longer than the oscillation period $1/\omega_{\text{eff}}$ of the membrane). (iii) the system dynamics should be stable when taking into account optical rigidity which is approximately equal to $G_0^2 \bar{c}^2/\omega_s$ for $\omega_m \ll \omega_s$.

Specifically, the SQL in condition (i), set by the first two terms in ΔN^2, gives $\gamma_c \gamma_d / G_{\text{eff}}^2 \lesssim 1$, or equivalently $(\gamma_c \gamma_d / G_0^2) \lesssim \Lambda^2$. If we neglect the optical spring effect ($\Lambda = 1$), we simply recover Eq. (9.8). A strong negative optical rigidity ($\omega_{\text{eff}} \ll \omega_m$, i.e. $\Lambda \gg 1$) can significantly enhance the effective coupling strength and ease the requirements on optomechanical properties. However, a small ω_{eff} also makes the system susceptible to the thermal noise. Taking account of all the above conditions, the optimal $\omega_{\text{eff}} = \omega_m \sqrt{\bar{n}_{\text{th}}/Q_m}$ with mechanical quality factor $Q_m \equiv \omega_m/\gamma_m$, and there is a nontrivial constraint on the thermal occupation number, namely $(\bar{n}_{\text{th}}/Q_m) \leq [G_0^2/(\omega_s \gamma_c)]^{2/3}$.

For a numerical estimate, we use a similar specification as given in Ref. [16], but assume a slightly higher mechanical quality factor Q_m, a lower environmental temperature T, and a lower input optical power I_0 such that all mentioned conditions are satisfied. The parameters are the following: $m = 50\,\text{pg}$, $\omega_m/(2\pi) = 10^5\,\text{Hz}$, $Q_m = 3.2 \times 10^7$, $\lambda = 532\,\text{nm}$, $L = 3\,\text{cm}$, $r_m = 0.9999$, $\mathcal{F} = 6 \times 10^5$, $T = 0.1\,\text{K}$ and $I_0 \approx 5\,\text{nW}$. The resulting resolution ΔN is shown in Fig. 9.2, and we are able to resolve single mechanical quantum when $\tau \approx 0.1\,\text{ms}$.

Even though we have been focusing on a double-sided setup where $t_1 \approx t_2$, the quantum limit also exists in the single-sided case originally proposed in Ref. [16]. Ideally, a single-sided setup consists of a totally-reflected end mirror and the vacuum fluctuations only enter from the front mirror. Therefore, the quantum noises inside the two sub-cavities have the same origin, but different optical path. Through

9.3 Coupled cavities 147

similar input-output calculations, we find that if laser detuning is equal to $\pm\omega_s$, the quantum noises destructively interfere with each other at low frequencies, due to the same mechanism studied in great details in Ref. [4], achieving an ideal QND measurement. However, in reality, the end mirror always has some finite transmission or optical loss which introduces uncorrelated vacuum fluctuations. As it turns out, the quantum limit is similar to Eq. (9.8), only with $\gamma_{c,d}$ replaced by the damping rate of two sub-cavities.

9.4 General Systems

Actually, the SQL obtained above applies to all schemes that attempt to probe mechanical energy quantization via parametric coupling. Let us consider n mechanical modes parametrically coupled with n' normal external modes, describable by the following Hamiltonian:

$$\hat{\mathcal{H}} = \sum_{\nu=1}^{n} \hbar\Omega_\nu(\hat{q}_\nu^2 + \hat{p}_\nu^2)/2 + \sum_{i=1}^{n'} \hbar\omega_i \hat{a}_i^\dagger \hat{a}_i \qquad (9.18)$$
$$+ \sum_{i,j=1}^{n'} \sum_{\nu=1}^{n} \hbar\chi_{ij\nu} \hat{q}_\nu(\hat{a}_i^\dagger \hat{a}_j + \hat{a}_j^\dagger \hat{a}_i) + \hat{\mathcal{H}}_{\text{ext}} + \hat{\mathcal{H}}_\xi \,.$$

Here Greek indices identify mechanical modes and Latin indices identify external modes; Ω_ν and ω_i are eigenfrequencies; $\hat{q}_\nu$, $\hat{p}_\nu$ are normalized positions and momenta; $\hat{a}_i$ are annihilation operators of the external degrees of freedom; $\chi_{ij\nu} = \chi_{ji\nu}$ are coupling constants. Similarly, we focus on the regime where $|\chi_{ij\nu}| \ll |\omega_i - \omega_j|$ (dispersive) and $\Omega_\nu \ll |\omega_i - \omega_j|$ (adiabatic), and obtain

$$\hat{\mathcal{H}} = \sum_{\nu=1}^{n} \hbar\Omega_\nu(\hat{q}_\nu^2 + \hat{p}_\nu^2)/2 + \sum_{i=1}^{n'} \hbar\omega_i' \hat{o}_i^\dagger \hat{o}_i + \hat{\mathcal{H}}_{\text{ext}} + \hat{\mathcal{H}}_\xi, \qquad (9.19)$$

where, up to $\chi_{ij\nu}^2/|\omega_i - \omega_j|^2$,

$$\omega_i' = \omega_i + \sum_\nu \chi_{ii\nu}\hat{q}_\nu + \sum_{j \neq i}\sum_\nu \frac{(\chi_{ij\nu}\hat{q}_\nu)^2}{\omega_i - \omega_j}. \qquad (9.20)$$

In order to have quadratic couplings between a pair of external and mechanical modes, $\hat{o}_1$ and $\hat{q}_1$ for instance, we require that $\chi_{11\nu} = 0$ and $\chi_{1i\nu} = \chi_{1i1}\delta_{1\nu}$, and then

$$\omega_1' = \omega_1 + \sum_{i \neq 1} \frac{\chi_{1i1}^2}{\omega_1 - \omega_i} \hat{q}_1^2. \qquad (9.21)$$

However, there still are linear couplings which originate from idle modes. This is because, up to $\chi_{ij\nu}/|\omega_i - \omega_j|$,

$$\hat{o}_i = \hat{a}_i + \sum_{j \neq i} \frac{\chi_{ij1}\hat{a}_j}{\omega_i - \omega_j} \hat{q}_1 \approx \hat{a}_i + \frac{\chi_{1i1}\bar{a}_1}{\omega_i - \omega_1} \hat{q}_1 (i \neq 1). \tag{9.22}$$

where $\hat{a}_1$ is replaced with its classical amplitude $\bar{a}_1$, for $\bar{a}_1 \gg \hat{a}_i$. From Eqs. (9.21) and (9.22), both linear and dispersive couplings are inversely proportional to $|\omega_i - \omega_1|$. Therefore, we only need to consider a tripartite system formed by $\hat{q}_1$, $\hat{o}_1$ and $\hat{o}_2$ which is the closest to $\hat{o}_1$ in frequency. The resulting Hamiltonian is identical to Eq. (9.2), and thus the same standard quantum limit applies.

9.5 Conclusions

We have demonstrated the existence of standard quantum limit for probing mechanical energy quantization in general systems where mechanical modes parametrically interact with optical or electrical degrees of freedom. This work will shed light on choosing the appropriate parameters for experimental realizations.

References

1. M. Bhattacharya, P. Meystre, Trapping and cooling a mirror to its quantum mechanical ground state. Phys. Rev. Lett. **99**, 073601 (2007)
2. A.A. Clerk, D.W. Utami, Using a qubit to measure photon-number statistics of a driven thermal oscillator. Phys. Rev. A **75**, 042302 (2007)
3. A.A. Clerk, M.H. Devoret, S.M. Girvin, F. Marquardt, R.J. Schoelkopf (2008) Introduction to quantum noise, measurement and amplification, arXiv:0810.4729
4. F. Elste, S.M. Girvin, A.A. Clerk, Quantum noise interference and backaction cooling in cavity nanomechanics. Phys. Rev. Lett. **102**, 207209 (2009)
5. C. Gardiner, P. Zoller, *Quantum Noise* (Springer, Berlin, 2004)
6. K. Jacobs, P. Lougovski, M. Blencowe, Continuous measurement of the energy eigenstates of a nanomechanical resonator without a nondemolition probe. Phys. Rev. Lett. **98**, 147201 (2007)
7. A.M. Jayich, J.C. Sankey, B.M. Zwickl, C. Yang, J.D. Thompson, S.M. Girvin, A.A. Clerk, F. Marquardt, J.G.E. Harris, Dispersive optomechanics: a membrane inside a cavity. New J Phys **10**(9), 095008 (2008)
8. S. Mancini, D. Vitali, P. Tombesi, Scheme for teleportation of quantum states onto a mechanical resonator. Phys. Rev. Lett. **90**, 137901 (2003)
9. F. Marquardt, J.P. Chen, A.A. Clerk, S.M. Girvin, Quantum theory of cavity-assisted sideband cooling of mechanical motion. Phys. Rev. Lett. **99**, 093902 (2007)
10. I. Martin, W.H. Zurek, Measurement of energy eigenstates by a slow detector. Phys. Rev. Lett. **98**, 120401 (2007)
11. H. Miao, S. Danilishin, T. Corbitt, Y. Chen, Standard quantum limit for probing mechanical energy quantization. Phys. Rev. Lett. **103**, 100402 (2009)

12. H. Mueller-Ebhardt, H. Rehbein, R. Schnabel, K. Danzmann, Y. Chen, Entanglement of Macroscopic Test Masses and the Standard Quantum Limit in Laser Interferometry. Phys. Rev. Lett. **100**, 013601 (2008)
13. A. Naik, O. Buu, M.D. LaHaye, A.D. Armour, A.A. Clerk, M.P. Blencowe, K.C. Schwab, Cooling a nanomechanical resonator with quantum back-action. Nature **443**, 193–196 (2006)
14. D.H. Santamore, A.C. Doherty, M.C. Cross, Quantum nondemolition measurement of Fock states of mesoscopic mechanical oscillators. Phys. Rev. B **70**, 144301 (2004)
15. D.I. Schuster, A.A. Houck, J.A. Schreier, A. Wallraff, J.M. Gambetta, A. Blais, L. Frunzio, J. Majer, B. Johnson, M.H. Devoret, S.M. Girvin, R.J. Schoelkopf, Resolving photon number states in a superconducting circuit. Nature **445**, 515–518 (2007)
16. J.D. Thompson, B.M. Zwickl, A.M. Jayich, F. Marquardt, S.M. Girvin, J.G.E. Harris, Strong dispersive coupling of a high-finesse cavity to a micromechanical membrane. Nature **452**, 72–75 (2008)
17. D. Vitali, S. Gigan, A. Ferreira, H.R. Böhm, P. Tombesi, A. Guerreiro, V. Vedral, A. Zeilinger, M. Aspelmeyer, Optomechanical entanglement between a movable mirror and a cavity field. Phys. Rev. Lett. **98**, 030405 (2007)

Chapter 10
State Preparation: Non-Gaussian Quantum State

10.1 Preface

As we conclude from the previous chapter, in order to create non-Gaussian quantum states, a nonlinear optomechanical coupling is generally required. This is rather challenging to achieve, especially when the mass of the mechanical oscillator is large and the frequency is low. In this chapter, we propose a protocol for coherently transferring non-Gaussian quantum states from the optical field to a mechanical oscillator, which does not require a nonlinear coupling in the optomechanical system. We demonstrate its experimental feasibility in both future gravitational-wave detectors and table-top optomechanical devices. This work not only outlines a feasible way to investigate non-classicality in macroscopic optomechanical systems, but also presents a new and elegant approach for solving non-Markovian open quantum dynamics in general linear systems. This is a joint research effort by Farid Khalili, Stefan Danilishin, Helge Müller-Edhardt, Huan Yang, Yanbei, and myself. It has been published in Phys. Rev. Letts. 105, 070403 (2010).

10.2 Introduction

Recently, there have been intensive experimental and theoretical studies on the investigation of quantum behaviors of macroscopic mechanical oscillators in optomechanical devices [20]. These activities are motivated by: (i) the necessity to achieve and go beyond the *Standard Quantum Limit* (SQL) for high-precision measurements with mechanical probes [2], (ii) the testing and interpretation of quantum theory with macroscopic degrees of freedom [17], and (iii) quantum information processing with optomechanical devices [20]. Non-Gaussian quantum states, such as Fock states, lie in the heart of all these endeavors [1, 3]. A 6 GHz micromechanical oscillator has been recently prepared in a single-quantum state by first cooling it down to the ground state with a dilution refrigeration, and then later swapping the quantum state between

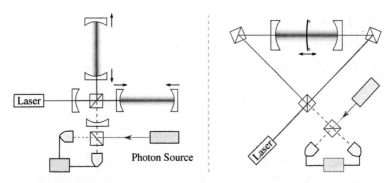

Fig. 10.1 Possible experimental schemes: **a** an interferometric gravitational-wave detector with kg-scale test masses (*left*) [10]. and **b** A tabletop coupled cavity scheme with a ng-scale membrane (*right*) [30]. From Ref. [14]

it and a superconducting qubit [28]. This is possible due to the intrinsic nonlinearity in the qubit system [4, 11]. For optomechanical system, to achieve nonlinearity in the quantum regime, the zero-point displacement $x_q = \sqrt{\hbar/(2m\omega_m)}$ of the oscillator with mass m and frequency ω_m is required to be comparable to the cavity linear dynamical range given by the optical wavelength λ divided by the cavity finesse $\mathcal{F}$ [15, 22, 24, 30]:

$$\lambda/(\mathcal{F} x_q) \lesssim 1. \tag{10.1}$$

Since, in a typical setup, $\lambda \sim 10^{-6}$ m and $\mathcal{F} \lesssim 10^6$, this gives $x_q \gtrsim 10^{-12}$ m, which is rather challenging to achieve.

Here, we propose a protocol for preparations of non-Gaussian quantum states with optomechanical devices, which *does not require an optomechanical nonlinearity*. The idea is to inject a non-Gaussian optical state, e.g., a single-photon pulse created by a cavity QED process [13, 16, 23], into the dark port of an interferometric optomechanical device as shown schematically in Fig. 10.1. The radiation-pressure force of the single photon on the mechanical oscillator is coherently amplified by the classical pumping from the bright port, and, as we will show, the qualitative requirement for preparing a non-Gaussian state becomes:

$$\lambda/(\mathcal{F} x_q) \lesssim \sqrt{N_\gamma}. \tag{10.2}$$

Here $N_\gamma = I_0 \tau/(\hbar \omega_0)$ (I_0 is the laser power into the bright port and ω_0 the frequency) is the number of pumping photons within the duration τ of the single-photon pulse, and we gain a significant factor of $\sqrt{N_\gamma}$ compared with Eq. (10.1), making this scheme experimentally achievable.

To motivate experiments and make correct predictions, it is crucial to understand (i) *the dynamics*: how the mechanical oscillator interacts with the single-photon pulse, and (ii) *the conditional process*: how the continuous measurement affects the final quantum state. For the dynamics, we will study the full quantum dynamics without using either the rotating-wave approximation [18], or the three-mode approach [18, 29], because the interaction timescale will be shorter than one mechanical oscillation period, in order to minimize the thermal decoherence effect–a strong optomechanical coupling. For the conditional process, the non-trivial quantum correlations at different times in the photon pulse make the open quantum dynamics highly non-Markovian, which does not allow a transparent study with the standard *Stochastic-Master-Equation* (SME) approach [5, 6, 7, 9, 26]. We develop a path-integral approach. This solves the non-Markovian dynamics elegantly and gives an explicit expression for the final quantum state of the mechanical oscillator, which is also valid for general linear systems.

The outline of this chapter is as follows: in Sect. 10.3, we will make an order-of-magnitude estimate of the experiment requirements of such a protocol by considering a simple case. In Sect. 10.4, we will present the path-integral based approach to treat non-Gaussian state preparation in general optomechanical systems, and an explicit relation between the optical state and the oscillator state is obtained. In Sect. 10.5, we will apply this result to the single-photon case, and find justifications for the previous order-of-magnitude estimate. Finally, we will summarize our main results in Sect. 10.6. In the Appendix, there are further details of the equations and concepts that we have introduced in the main sections.

10.3 Order-of-Magnitude Estimate

The model of such optomechanical devices is shown in Fig. 10.2. The oscillator position $\hat{x}$ is coupled to a thermal bath and also the cavity mode $\hat{a}$ (mediated by radiation pressure) which in turn interacts with ingoing $\hat{a}_{in}$ and outgoing $\hat{a}_{out}$ optical fields. To gain a qualitative picture, we first make an order-of-magnitude estimate by considering a simple case in which the cavity bandwidth is large with $\hat{a}$ adiabatically eliminated and the oscillator is a free mass (frequency $\omega_m \sim 0$) with the thermal force ignored. The corresponding Heisenberg equations of motion read (refer to (Appendix 10.7.1) for more details):

$$\dot{\hat{x}}(t) = \hat{p}(t)/m, \quad \dot{\hat{p}}(t) = \alpha \hat{a}_1(t), \tag{10.3}$$
$$\hat{b}_1(t) = \hat{a}_1(t), \quad \hat{b}_2(t) = \hat{a}_2(t) + (\alpha/\hbar)\hat{x}(t). \tag{10.4}$$

Here, the coupling constant $\alpha \equiv 8\sqrt{2}(\mathcal{F}/\lambda)\sqrt{\hbar I_0/\omega_0}$; $\hat{a}_{1,2}$ ($\hat{b}_{1,2}$) are the amplitude and phase quadratures of the ingoing field $\hat{a}_{in}$ (outgoing field $\hat{a}_{out}$) with $\hat{a}_1 \equiv (\hat{a}_{in} + \hat{a}_{in}^{\dagger})/\sqrt{2}$ and $\hat{a}_2 \equiv (\hat{a}_{in} - \hat{a}_{in}^{\dagger})/(i\sqrt{2})$ (the same for the relation between $\hat{b}_{1,2}$ and $\hat{a}_{out}$).

Suppose, at $t = -\tau$, the oscillator is prepared in some initial quantum state $|\psi_m\rangle = \int_{-\infty}^{\infty} \psi_m(x)|x\rangle dx$ and is interacting with a single photon up to $t = 0$. With a short photon pulse (i.e., a short interaction duration), the oscillator position almost does not change, and we obtain:

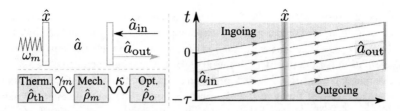

Fig. 10.2 A model of the optomechanical device shown in Fig. 10.1 (*upper left*), its spacetime diagram (*right*), and the couplings (*lower left*). Ingoing and outgoing rays (*tilted lines*) are placed on opposite sides of the oscillator world line (*vertical*) for clarity. The ingoing field contains a photon pulse, while the outgoing field—which contains the information of the oscillator motion—is measured continuously. From Ref. [14]

$$\hat{X}(0) = \hat{X}(-\tau), \quad \hat{P}(0) = \hat{P}(-\tau) + \kappa \hat{A}_1, \tag{10.5}$$

$$\hat{B}_1 = \hat{A}_1, \quad \hat{B}_2 = \hat{A}_2 + \kappa \hat{X}(0). \tag{10.6}$$

We have normalized the oscillator position and momentum by their zero-point uncertainties: $\hat{X} \equiv \hat{x}/x_q$ ($x_q \equiv \sqrt{\hbar/2m\omega_m}$) and $\hat{P} \equiv \hat{p}/p_q$ ($p_q \equiv \sqrt{\hbar m \omega_m/2}$); we have introduced $\hat{A}_j = \sqrt{2/\tau} \int_{-\tau}^{0} dt \hat{a}_j(t)$ ($j=1,2$) which has an uncertainty $\Delta \hat{A}_j = 1$ with $\Delta \hat{A} \equiv \langle \psi | (\hat{A} - \bar{A})^2 | \psi \rangle^{1/2}$; $\hat{B}_j = \sqrt{2/\tau} \int_{-\tau}^{0} dt \hat{b}_j(t)$; $\kappa \equiv \alpha \sqrt{2\tau} x_q / \hbar = 16 \sqrt{N_\gamma} \mathcal{F} x_q / \lambda$. These two equations are similar to those for studying the atom-light interaction in a quantum memory [12, 19]. They can be transformed back into an evolution operator: $\hat{U} = \exp[i\kappa \hat{A}_1 \hat{X}]$ in the Schrödinger picture. The quantum state of the system at $t = 0$ is simply $|\psi\rangle = \hat{U}|\psi_o\rangle|\psi_m\rangle$ with $|\psi_o\rangle$ the initial optical state. Given a measurement of $\hat{B}_2$ with a precise result y, the oscillator is projected into a conditional quantum state: $|\psi_m^c\rangle = \langle y|\hat{U}|\psi_o\rangle \psi_m\rangle$ which, in the coordinate representation $\psi_m^c(X) \equiv \langle X|\psi_m^c\rangle$, reads

$$\psi_m^c(X) = \psi_o(y - \kappa X)\psi_m(X) \tag{10.7}$$

—*the optical state is mapped onto the mechanical oscillator.* A significant mapping requires that $\psi_o(y-\kappa X)$ dominates over $\psi_m(X)$ in determining the profile of $\psi_m^c(X)$. In momentum space, this dictates that the momentum uncertainty induced by the optomechanical interaction should be large than that from the initial state $\psi_m(X)$, i.e., $\kappa \Delta \hat{A}_1 > \Delta \hat{P}(-\tau)$. Suppose the oscillator is initially in its quantum ground state with $\Delta \hat{P}(-\tau) = 1$. Since $\Delta \hat{A}_1 = 1$, this condition reads $\kappa > 1$, i.e., $\lambda/(\mathcal{F} x_q) < 16\sqrt{N_\gamma}$, which justifies Eq. (10.2).

In the above considerations, we have ignored the important thermal decoherence effect. In a real experiment, it is essential that the momentum fluctuations due to the thermal force within the interaction duration τ—$\Delta \hat{P}_{\text{th}} = (S_{FF}^{\text{th}} \tau)^{1/2}/p_q$—should be small compared with that from the optomechanical interaction, namely, $\Delta \hat{P}_{\text{th}} < \kappa$. In the high temperature limit, the force spectrum S_{FF}^{th} of the thermal force is $2m\omega_m k_B T/Q_m$ with T the environmental temperature and Q_m the mechanical quality factor. More explicitly, such a requirement reads

10.3 Order-of-Magnitude Estimate

Table 10.1 Possible experimental specifications

Parameters	λ (μm)	$\mathcal{F}$	m	$\omega_m/2\pi$ (Hz)	Q_m	T (K)	I_0	τ (ms)
large scale	1	6,000	4 kg	1	10^8	300	200 W	1
small scale	1	10^4	1 ng	10^5	10^7	4	0.2 μW	0.01

$$\lambda/(\mathcal{F}x_q)\sqrt{n_{\text{th}}/Q_m}\sqrt{\omega_m\tau} < 8\sqrt{N_\gamma}. \qquad (10.8)$$

with $n_{\text{th}} \equiv k_B T/(\hbar\omega_m)$ the thermal occupation number. *These two conditions* [cf. Eqs. (10.2) and (10.8)] *set the benchmarks for a successful non-Gaussian state-preparation experiment.* They can be satisfied with experimentally feasible specifications as shown in Table 10.1, in which the first row is close to that of a large-scale gravitational-wave detector and the second row to that of a small-scale optomechanical device in Ref. [30]. Such a qualitative picture will be justified by a rigorous treatment below.

10.4 General Formalism

For a quantitative study, we divide the entire process from $t = -\tau$ to $t = 0$ into N segments and later take the continuous limit. The n-th segment consists of: (i) *a free evolution,* which is described by an evolution operator: $\hat{\mathcal{U}}_n \equiv \exp[-i\hat{H}_n\tau/(N\hbar)]$ with $\hat{H}_n$ the system Hamiltonian at $t = (n-N)\tau/N$, and (ii) *a measurement of the outgoing field at a certain quadrature* $\hat{y} = \hat{b}_1\cos\theta + \hat{b}_2\sin\theta$, which is described by a projection operator: $\hat{\mathcal{P}}_n = \delta(\hat{y} - y_n)$ with y_n the measurement result of $\hat{y}$. After the entire process and conditioning on the measurement results $\mathbf{y} = (y_1, \ldots, y_N)$, the system is projected into a conditional quantum state:

$$\hat{\rho}^c(\mathbf{y}) = \hat{\mathcal{P}}_\mathbf{y}\hat{\rho}_i\hat{\mathcal{P}}_\mathbf{y}^\dagger/w(\mathbf{y}) \qquad (10.9)$$

with $\hat{\mathcal{P}}_\mathbf{y} \equiv \hat{\mathcal{P}}_N\hat{\mathcal{U}}_N \cdots \hat{\mathcal{P}}_1\hat{\mathcal{U}}_1$ and $w[\mathbf{y}] \equiv \text{Tr}_{\text{th},o,m}[\hat{\mathcal{P}}_\mathbf{y}\hat{\rho}_i\hat{\mathcal{P}}_\mathbf{y}^\dagger]$ the probability for obtaining measurement results $\mathbf{y}$. The initial density matrix $\hat{\rho}_i$ of the system is $\hat{\rho}_{\text{th}}(-\tau) \otimes \hat{\rho}_o(-\tau) \otimes \hat{\rho}_m(-\tau)$ with $\hat{\rho}_{\text{th}}$, $\hat{\rho}_o$, and $\hat{\rho}_m$ for the thermal bath, optical field, and mechanical oscillator, respectively.

The conditional quantum state of the mechanical oscillator $\hat{\rho}_m^c$ is obtained by tracing out the degrees of freedom of both the thermal heat bath and optical field, i.e.,

$$\hat{\rho}_m^c(\mathbf{y}) = \text{Tr}_{\text{th},o}[\hat{\rho}^c(\mathbf{y})] = \text{Tr}_{\text{th},o}[\hat{\mathcal{P}}_\mathbf{y}\hat{\rho}_i\hat{\mathcal{P}}_\mathbf{y}^\dagger]/w(\mathbf{y}). \qquad (10.10)$$

In the standard SME approach, such a trace operation is made right after each segment and this requires that these degrees of freedom at different segments to be not correlated. However, it is not satisfied here, due to quantum correlations among different segments (non-Markovian) arising from the non-trivial initial optical state.

We apply a different approach based upon a path integral. By using the facts that: $\hat{\mathcal{P}}_y = \hat{U}(\tau)\hat{\mathcal{P}}_N^H\hat{\mathcal{P}}_{N-1}^H \ldots \hat{\mathcal{P}}_1^H \equiv \hat{U}(\tau)\hat{\mathcal{P}}_y^H$, where $\hat{U}(\tau) \equiv \prod_{n=1}^{N}\hat{U}_n$ (time-ordered) and $\hat{\mathcal{P}}_n^H \equiv \delta(\hat{y}_n - y_n)$ with $\hat{y}_n \equiv \hat{U}^\dagger(\frac{n\tau}{N})\hat{y}\hat{U}(\frac{n\tau}{N})$, and the optical quadrature at different times commute: $[\hat{y}_n, \hat{y}_{n'}] = 0$, we obtain $\hat{\mathcal{P}}_y^H = \prod_{n=1}^{N}\delta(\hat{y}_n - y_n) = \int \frac{d^N\xi}{(2\pi)^N} \exp[i\sum_{n=1}^{N}\xi_n(\hat{y}_n - y_n)]$. In the continuous limit $N \to \infty$, the total projection operator can be rewritten as a path integral:

$$\hat{\mathcal{P}}_y = \hat{U}(\tau)\int \mathcal{D}[\xi] \exp\left\{i\int_{-\tau}^{0} dt\xi(t)[\hat{y}(t) - y(t)]\right\}, \tag{10.11}$$

which allows us to take the *entire measurement history*, and trace out the optical field in a single step, instead of sequentially as in the SME approach.

To obtain an explicit expression for the conditional quantum state of the mechanical oscillator, i.e., its Wigner function, we evaluate the generating function:

$$\mathcal{J}[\alpha_x, \alpha_p; \mathbf{y}] = \text{Tr}_m[e^{i\alpha_x\hat{x}+i\alpha_p\hat{p}}\hat{\rho}_m^c(\mathbf{y})], \tag{10.12}$$

which is related to the Wigner function by $W[x, p; \mathbf{y}] = \int \frac{d^2\alpha}{(2\pi)^2}e^{-i(\alpha_x x+\alpha_p p)}\mathcal{J}[\alpha_x, \alpha_p; \mathbf{y}]$.

From the facts that: $\hat{U}(\tau)^\dagger\hat{x}\hat{U}(\tau) = \hat{x}(0)$ and $[\hat{x}(0), \hat{y}(t)] = 0 (t < 0)$ (also true for $\hat{p}$), and the property $\hat{\mathcal{P}}_y^\dagger\hat{\mathcal{P}}_y = \hat{\mathcal{P}}_y$, we obtain

$$\mathcal{J}[\alpha; \mathbf{y}] = \text{Tr}_{th,m,o}[e^{i\alpha\mathbf{x}_0'}\hat{\mathcal{P}}_y^H\hat{\rho}_i]/w(\mathbf{y}), \tag{10.13}$$

where vectors $\alpha \equiv (\alpha_x, \alpha_p)$, $\hat{\mathbf{x}}_0 \equiv (\hat{x}(0), \hat{p}(0))$, and superscript $'$ denotes transpose. To move forward, we need to specify the initial density matrix $\hat{\rho}_i$ of the system. For the thermal bath in thermal equilibrium at temperature T, $\hat{\rho}_{th}(-\tau) = e^{-\hat{H}_{th}/k_BT}/\text{Tr}[e^{-\hat{H}_{th}/k_BT}]$. For the optical field, we consider an arbitrary spatial profile $f(x/c)$ for the photon pulse of which the creation operator reads $\hat{\Gamma}^\dagger \equiv \int_{-\tau}^{0} dt f^*(t)\hat{a}_{in}^\dagger(t)$. In the P-presentation, a general initial state of such a mode can be written as $\hat{\rho}_o(-\tau) = \int d^2\zeta P(\zeta)|\zeta\rangle\langle\zeta|$ with the vector $\zeta \equiv (\Re[\zeta], \Im[\zeta])$ and $|\zeta\rangle \equiv \exp[\zeta\hat{\Gamma}^\dagger - \zeta^*\hat{\Gamma}]|0\rangle$. Since the timescale of the photon profile $f(t)$ will automatically set the interaction duration, we can extend $-\tau$ to $-\infty$ which is equivalent to turning on the optomechanical interaction adiabatically. In this case, the initial state of the oscillator-coupled to the thermal bath-decays away before the optomechanical interaction starts, and thus does not influence $\hat{\rho}_m^c$.

By substituting in the initial state and Baker-Campbell-Hausdorff formula, the generating function becomes:

$$\mathcal{J} = \frac{1}{w(\mathbf{y})}\int d^2\zeta \mathcal{D}[\xi]e^{i[\zeta^*\hat{\Gamma} - \zeta\hat{\Gamma}^\dagger, \hat{B}]}\langle 0|e^{i\hat{B}}|0\rangle P(\zeta), \tag{10.14}$$

where $\hat{B} \equiv \alpha\hat{\mathbf{x}}_0' + \int_{-\infty}^{0} dt\xi(t)[\hat{y}(t) - y(t)]$. Further evaluation of $\mathcal{J}$ requires us to manipulate the statistics of the measured optical quadrature $\hat{y}(t)$ and the oscillator

10.4 General Formalism

motion $\hat{\mathbf{x}}_0$. We apply the tools introduced in Ref. [27] (refer to Appendix 10.7.2 for some details) to (i) *simplify the statistics of $\hat{y}(t)$*, while maintaining its full information by causally whitening it into $\hat{z}(t)$ such that $\langle\hat{z}(t)\hat{z}(t')\rangle = \delta(t-t')$, and (ii) *separate $\hat{\mathbf{x}}_0$ into a quantum part $\hat{\mathbf{R}}$ and a classical part* which can be inferred from $\hat{z}$ by using the optimal Wiener filter $\mathbf{K} - \hat{\mathbf{x}}_0 \equiv \hat{\mathbf{R}} + \int_{-\infty}^{0} dt\, \mathbf{K}(-t)\hat{z}(t)$-such that $\hat{\mathbf{R}}$ is not correlated with $\hat{z}(t)$, namely $\langle 0|\hat{\mathbf{R}}\hat{z}(t)|0\rangle = 0$. With these tools, the path integral can be completed, and it gives

$$\mathcal{J} = \frac{1}{w(\mathbf{y})}\int d^2\zeta\, e^{-|\alpha\mathbb{V}_c\alpha' + \|z - 2\zeta L'\|^2|/2 + i\alpha\mathbf{x}_{\zeta'}} P(\zeta). \tag{10.15}$$

Here $\mathbb{V}_c \equiv \langle 0|\hat{\mathbf{R}}'\hat{\mathbf{R}}|0\rangle$ with $\hat{\mathbf{R}} = (\hat{R}_x, \hat{R}_p)$; the modulus of a function: $\|g\|^2 \equiv \int_{-\infty}^{0} g(t)g^*(t)dt$; $\mathbf{x}_\zeta \equiv \mathbf{x}_c + \zeta^*\gamma + \zeta\gamma^*$; $\mathbf{x}_c \equiv (x_c, p_c) = \int_{-\infty}^{0} dt\, \mathbf{K}(-t)z(t)$ with $z(t)$ measured results of $\hat{z}(t)$ and $\mathbf{K} = (K_x, K_p)$; $\gamma \equiv [\hat{\Gamma}, \hat{\mathbf{R}}]$ and $L \equiv (\Re[L], \Im[L])$ with $L(t) \equiv [\hat{\Gamma}, \hat{z}(t)]$, which characterize the contribution of the photon $\hat{\Gamma}$ to both the oscillator motion $\hat{\mathbf{R}}$ and the output field $\hat{z}$, and determine the efficiency of the state transfer.

Finally, the Wigner function for the quantum state of the oscillator reads (the normalization factor is ignored):

$$W = \int d^2\zeta\, e^{-[(\mathbf{x}-\mathbf{x}_\zeta)\mathbb{V}_c^{-1}(\mathbf{x}-\mathbf{x}_\zeta)' + \|z - 2\zeta L'\|^2]/2} P(\zeta) \tag{10.16}$$

with $\chi \equiv \mathbf{x} - \mathbf{x}_\zeta$. *This formula directly relates the optical state to the resulting state of the mechanical oscillator.* Since no specific Hamiltonian is assumed in deriving it, it is valid for general linear quantum dynamics. For cavity-assisted optomechanical systems, γ, $\mathbb{V}_c$, $\mathbf{K}$ and L can be obtained from the standard input-output relations in Refs. [8, 21, 31] by using the formalism in Ref. [27]. The state transfer efficiency can be measured quantitatively by the fidelity defined as $F \equiv \text{Tr}[\hat{\rho}_m^c\hat{\rho}_o]$ which is equal to the overlapping between two quantum states [12, 19] (cf. Appendix 10.7.3).

10.5 Single-Photon Case

As an example, we consider the simplest case where the optical field is in a single-photon state with $\hat{\rho}_o = |1\rangle\langle 1|$ and $P(\zeta) = e^{|\zeta|^2}\partial^2\delta^{(2)}(\zeta)/\partial\zeta\partial\zeta^*$. From Eq. (10.16), the normalized Wigner function reads:

$$W = \frac{1 - \gamma\mathbb{V}_c^{-1}\gamma^\dagger - \|L\|^2 + |\gamma\mathbb{V}_c^{-1}\delta\mathbf{x}' + Z|^2}{2\pi\sqrt{\det\mathbb{V}_c}(1 - \|L\|^2 + |Z|^2)} e^{-\delta\mathbf{x}\mathbb{V}_c^{-1}\delta\mathbf{x}'/2}, \tag{10.17}$$

where $\delta\mathbf{x} \equiv \mathbf{x} - \mathbf{x}_c$ and $Z \equiv \int_{-\infty}^{0} dt\, z(t)L(t)$. Since the measurement results $z(t)$ only appear in the above Wigner function in terms of an integral, i.e., Z, the conditional

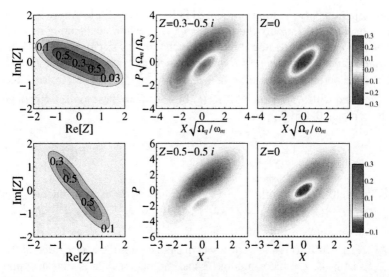

Fig. 10.3 Distributions of measurement results (*left* panels) and the corresponding Wigner function of the oscillator given the most probable measurement result (*middle* panels) and less probable result but with a significant non-Gaussianity (*right* panels). The upper panels show the case with a specification close to a gravitational-wave detector, and the lower panels close to a small-scale optomechanical device as listed in Table 10.1. We have used normalized coordinates (with respect to x_q and p_q) and introduced $\Omega_q \equiv \sqrt{\hbar m/\alpha^2}$. From Ref. [14]

process is easy to study and the random vector $\mathbf{Z} = (\Re[Z], \Im[Z])$ follows a two-dimensional distribution:

$$w[\mathbf{Z}] = \frac{1 - \|L\|^2 + \mathbf{Z}\mathbf{Z}'}{2\pi\sqrt{\det \mathbb{V}_L}} e^{-\mathbf{Z}\mathbb{V}_L^{-1}\mathbf{Z}'/2} \tag{10.18}$$

with $\mathbb{V}_L \equiv \int_{-\infty}^{0} dt\, \mathbf{L}'\mathbf{L}$.

With Eq. (10.17), we can justify the previous order-of-magnitude estimate by using the same specifications listed in Table 10.1. As an example, we assume a photon profile of $f(t) = \sqrt{2\gamma_f} e^{(\gamma_f + i\omega_f)t}$. The resulting Wigner functions of the mechanical oscillator are shown in Fig. 10.3. In the case of an advanced gravitational-wave detector, $\omega_f/2\pi = \gamma_f/2\pi = 70\,\text{Hz}$, and the state transfer fidelity $F = 0.58$ ($Z = 0.3 - 0.5i$) and $F = 0.95$ ($Z = 0$); in the case of a small-scale device, $\omega_f/\omega_m = 0.1$, $\gamma_f/\omega_m = 0.3$, and the corresponding $F = 0.34$ ($Z = 0.5 - 0.5i$) and $F = 0.56$ ($Z = 0$). In both cases, the Wigner function has negative regions–a unique quantum feature. The prepared non-Gaussian quantum state can be independently verified using the quantum tomography protocol proposed in Ref. [25] which allows us to reconstruct the quantum state with a sub-Heisenberg accuracy. This is essential for revealing these nonclassical negative regions.

10.6 Conclusions

We have outlined an experimental protocol for creating non-Gaussian quantum states in a macroscopic mechanical oscillator with optomechanical interactions. The radiation-pressure induced by the photon pulse is coherently amplified, and this allows us to transfer the optical state to the mechanical oscillator. Starting from an order-of-magnitude estimate, we have convinced ourself that this protocol is feasible for both future gravitational-wave detectors and small-scale table-top experiments. This has been confirmed by a more rigorous treatment in which a path integral is constructed for the measurement process. Such a path-integral-based approach provides an elegant treatment of the non-Markovian conditional dynamics in an open quantum system, and it is valid for general linear continuous measurements.

10.7 Appendix

10.7.1 Optomechanical Dynamics

In this section, we will briefly review the dynamics of a typical cavity-assisted optomechanical system, in order to justify some of the equations in the main text. The Hamiltonian for such an optomechanical system (shown schematically in Fig. 10.2) can be written as (cf. Refs. [8, 21, 31])

$$\hat{\mathcal{H}} = \frac{\hat{p}^2}{2m} + \frac{1}{2}m\omega_m^2\hat{x}^2 + \hbar\omega_c\hat{a}^\dagger\hat{a} + \hbar G_0\hat{x}\hat{a}^\dagger\hat{a} \\ + i\hbar\sqrt{2\gamma}(\hat{a}_{\text{in}}e^{-i\omega_0 t}\hat{a}^\dagger - \text{H.c.}) + \hat{H}_\gamma + \hat{H}_{\gamma_m}. \quad (10.19)$$

Here, $\hat{x}$ and $\hat{p}$ are the position and momentum operators for the oscillator; $\hat{a}$ is the annihilation operator for the cavity mode; $G_0 \equiv \omega_m/L$ is the optomechanical coupling constant with L the cavity length; γ is the cavity bandwidth; $\hat{H}_\gamma$ and $\hat{H}_{\gamma_m}$ describe the dissipation mechanism of the cavity mode and the mechanical oscillator, respectively.

In the rotating frame at the laser frequency ω_0, the above Hamiltonian leads to the following standard Langevin equations for the mechanical oscillator:

$$\dot{\hat{x}}(t) = \frac{\hat{p}(t)}{m}, \quad (10.20)$$

$$\dot{\hat{p}}(t) = -\gamma_m \hat{p}(t) - m\omega_m^2 \hat{x}(t) + \hbar G_0 \hat{a}^\dagger(t)\hat{a}(t) + \hat{F}_{\text{th}}(t), \quad (10.21)$$

with $\hat{F}_{\text{th}}$ the thermal force noise, and for the cavity mode:

$$\dot{\hat{a}}(t) + (\gamma - i\Delta)\hat{a}(t) = iG_0\hat{a}(t)\hat{x}(t) + \sqrt{2\gamma}\hat{a}_{\text{in}}(t), \quad (10.22)$$

with $\Delta \equiv \omega_0 - \omega_c$. The standard input-output relation for the cavity mode is given by (refer to Ref. [7])

$$\hat{a}_{\text{in}}(t) + \hat{a}_{\text{out}}(t) = \sqrt{2\gamma}\hat{a}(t). \tag{10.23}$$

Since the cavity mode is coherently driven with a laser, the above equation can be linearized by replacing every quantity $\hat{o}$ with a sum of the classical steady part $\bar{o}$ and a perturbed part. The equations of motion for the perturbed parts of the oscillator read

$$\dot{\hat{x}}(t) = \frac{\hat{p}(t)}{m}, \tag{10.24}$$

$$\dot{\hat{p}}(t) = -\gamma_m \hat{p}(t) - m\omega_m^2 \hat{x}(t) + \bar{G}_0 \hat{a}_1(t) + \hat{F}_{\text{th}}(t), \tag{10.25}$$

where the amplitude quadrature $\hat{a}_1(t) \equiv [\hat{a}(t) + \hat{a}^\dagger(t)]/\sqrt{2}$ and $\bar{G}_0 \equiv \sqrt{2\hbar}G_0\bar{a}$. Similarly, for the cavity mode,

$$\dot{\hat{a}}(t) + (\gamma - i\Delta)\hat{a}(t) = i\frac{\bar{G}_0}{\sqrt{2\hbar}}\hat{x}(t) + \sqrt{2\gamma}\hat{a}_{\text{in}}(t). \tag{10.26}$$

In the limit of a large cavity bandwidth considered in the order-of-magnitude estimate (refer to Sect. 10.3), the time dependence of the cavity mode can be adiabatically eliminated and we have

$$\hat{a}(t) \approx i\frac{\bar{G}_0}{\sqrt{2\hbar}\gamma}\hat{x}(t) + \sqrt{\frac{2}{\gamma}}\hat{a}_{\text{in}}(t). \tag{10.27}$$

Therefore, from Eq. (10.23),

$$\hat{a}_{\text{out}}(t) = \hat{a}_{\text{in}}(t) + i\frac{\alpha}{\sqrt{2\hbar}}\hat{x}(t). \tag{10.28}$$

with $\alpha \equiv \sqrt{2}\bar{G}_0/\sqrt{\gamma}$. By defining the output amplitude and phase quadratures as

$$\hat{b}_1 = \frac{\hat{a}_{\text{out}} + \hat{a}_{\text{out}}^\dagger}{\sqrt{2}}, \quad \hat{b}_2 = \frac{\hat{a}_{\text{out}} - \hat{a}_{\text{out}}^\dagger}{i\sqrt{2}}, \tag{10.29}$$

we recover what has been shown in Eq. (10.4).

10.7.2 Causal whitening and Wiener Filter

In this section, we will briefly introduce the concepts of causal whitening and Wiener filtering techniques applied in this paper (One can refer to Ref. [27] for more details). They are implemented extensively in the classical signal filtering.

10.7 Appendix

Here, the reason why these classical techniques can be applied lies in following fact: In a linear continuous measurement, the degrees of freedom of the measurement output $\hat{y}(t)$ at different times commute with each other [2], i.e.,

$$[\hat{y}(t), \hat{y}(t')] = 0. \tag{10.30}$$

This basically means that in principle, they can be simultaneously measured with arbitrarily high accuracy without imposing any limit. Therefore, they can be treated just as classical entities, and classical filtering techniques apply.

Causal whitening—Causal whitening is a powerful tool for simplifying the statistic of a random variable (the measurement output in this context), while maintaining its complete information. Mathematically, given the spectrum $S_{yy}(\Omega)$ of the output $\hat{y}(t)$, we can factorize it as:

$$S_{yy}(\Omega) = \phi_+(\Omega)\phi_-(\Omega), \tag{10.31}$$

such that ϕ_+ (ϕ_-) and its inverse are analytical functions in the upper- (lower-) half complex plane, and $\phi_+^* = \phi_-$. The causally-whitened output in the frequency domain is defined as

$$\hat{z}(\Omega) \equiv \frac{\hat{y}(\Omega)}{\phi_+(\Omega)}. \tag{10.32}$$

Since

$$\langle \hat{z}(\Omega)\hat{z}(\Omega') \rangle = 2\pi \frac{S_{yy}}{\phi_+\phi_-} \delta(\Omega - \Omega') = 2\pi \delta(\Omega - \Omega'), \tag{10.33}$$

the corresponding correlation function is:

$$\langle \hat{z}(t)\hat{z}(t') \rangle = \delta(t - t'), \tag{10.34}$$

which corresponds to a white noise with no correlations at different times. This not only simplifies the statistics, because $\hat{z}$ is uniquely defined from $\hat{y}$, it also possesses the same amount of information concerning the motion of the mechanical oscillator.

Wiener filter—A Wiener filter is the optimal filter satisfying the least mean-square error criterion. Given a random variable $\hat{x}$ (here the oscillator position), we can extract a maximal amount of information about $\hat{x}(0)$ from the measurement data $\hat{y}(t)$ (from $-\infty$ to 0) with the Wiener filter $K_x(t)$. The conditional mean of $\hat{x}(0)$ is

$$\hat{x}^{\text{cond}}(0) = \int_{-\infty}^{0} dt\, K_x(-t)\hat{y}(t). \tag{10.35}$$

The corresponding error $\hat{R}_x(t) = \hat{x}(0) - \hat{x}^{\text{cond}}(0)$ defines the remaining uncertainty that we cannot learn from $\hat{y}(t)$. Mathematically, this dictates that such an error is not correlated with $\hat{y}$ and is orthogonal to the space defined by the measurement results, namely,

$$\langle \hat{R}_x(0)\hat{y}(t)\rangle = 0. \tag{10.36}$$

Therefore, the decomposition that we applied in this paper and also Ref. [27]:

$$\hat{x}(0) = \int_{-\infty}^{0} dt\, K_x(-t)\hat{y}(t) + \hat{R}_x(0) \tag{10.37}$$

is very useful in separating the statistical dependence of $\hat{x}$ on the measurement $\hat{y}$ and facilitates the analysis of the conditional dynamics.

The Wiener filter can be obtained using the standard Wiener-Hopf method, and its frequency domain representation is:

$$K_x(\Omega) = \frac{1}{\phi_+(\Omega)}\left[\frac{S_{xy}(\Omega)}{\phi_-(\Omega)}\right]_+, \tag{10.38}$$

where S_{xy} is the cross-correlation between $\hat{x}$ and $\hat{y}$, and $[f(\Omega)]_+$ means taking the component of $f(\Omega)$ that is analytical in the upper-half complex plane.

10.7.3 State Transfer Fidelity

To quantify the state transfer, we follow Refs. [12, 19] by defining the fidelity from the overlap between the two Wigner functions of the prepared oscillator state $W_m(X, P)$ and the target state $W_{\text{tag}}(X, P)$:

$$F \equiv 2\pi \int_{-\infty}^{\infty} dX \int_{-\infty}^{\infty} dP\, W_m(X, P) W_{\text{tag}}(X, P). \tag{10.39}$$

Depending on the situation, X and P can be normalized with respect to either the zero-point uncertainty x_q and p_q or $x_q\sqrt{\omega_m/\Omega_q}$, and $p_q\sqrt{\Omega_q/\omega_m}$.

Since the center of the prepared state is given by x_c and p_c, we need to shift it to the center to compare with the target state, and this will not introduce any statistical difference. In addition, the prepared state is a squeezed state defined by $\mathbb{V}_c$. To properly evaluate the overlap, we will apply the well-known *Bogoliubov transformation* to the coordinates of the prepared state:

$$\hat{X}' = \hat{X}(\sinh\beta + \cosh\beta\cos 2\phi) - \hat{P}\cosh\beta\sin 2\phi, \tag{10.40}$$

$$\hat{P}' = \hat{P}(\sinh\beta + \cosh\beta\cos 2\phi) + \hat{X}\cosh\beta\sin 2\phi. \tag{10.41}$$

By choosing an appropriate set of squeezing factor β and rotation angle ϕ, the overlap with the target state can be maximized. Therefore, a properly modified definition for the fidelity should be:

$$F' \equiv \max[F, \{\beta, \phi\}]. \tag{10.42}$$

10.7 Appendix

In the case of the single-photon injection, the Wigner function of the target mechanical state is simply:

$$W_{\text{tag}}(X, P) = \frac{X^2 + P^2 - 1}{2\pi} \exp\left[-\frac{X^2 + P^2}{2}\right]. \tag{10.43}$$

References

1. J.S. Bell, *Speakable and Unspeakable in Quantum Mechanics* (Cambridge University Press, Cambridge, 1987)
2. V.B. Braginsky, F.Y. Khalili, *Quantum Measurement* (Cambridge University Press, Cambridge, 1992)
3. S.L. Braunstein, van P. Loock, Quantum information with continuous variables. Rev. Mod. Phys. **77**, 513 (2005)
4. A.A. Clerk, D.W. Utami, Using a qubit to measure photon-number statistics of a driven thermal oscillator. Phys. Rev. A **75**, 042302 (2007)
5. A.C. Doherty, K. Jacobs, Feedback control of quantum systems using continuous state estimation. Phys. Rev. A **60**, 2700 (1999)
6. A.C. Doherty, S.M. Tan, A.S. Parkins, D.F. Walls, State determination in continuous measurement. Phys. Rev. A **60**, 2380 (1999)
7. C. Gardiner, P. Zoller, *Quantum Noise* (Springer, Berlin, 2004)
8. C. Genes, D. Vitali, P. Tombesi, S. Gigan, M. Aspelmeyer, Groundstate cooling of a micromechanical oscillator: comparing cold damping and cavity-assisted cooling schemes. Phys. Rev. A **77**, 033804 (2008)
9. A. Hopkins, K. Jacobs, S. Habib, K. Schwab, Feedback cooling of a nanomechanical resonator. Phys. Rev. B **68**, 235328 (2003)
10. http://www.ligo.caltech.edu/advLIGO
11. K. Jacobs, P. Lougovski, M. Blencowe, Continuous measurement of the energy eigenstates of a nanomechanical resonator without a nondemolition probe. Phys. Rev. Lett. **98**, 147201 (2007)
12. B. Julsgaard, J. Sherson, J.I. Cirac, J. Fiurasek, E.S. Polzik, Experimental demonstration of quantum memory for light. Nature **432**, 482–486 (2004)
13. M. Keller, B. Lange, K. Hayasaka, W. Lange, H. Walther, Continuous generation of single photons with controlled waveform in an ion-trap cavity system. Nature **431**, 1075–1078 (2004)
14. F. Khalili, S. Danilishin, H. Miao, H. Mueller-Ebhardt, H. Yang, Y. Chen, Preparing a mechanical oscillator in non-Gaussian quantum states. submitted to PRL, 2010
15. D. Kleckner, I. Pikovski, E. Jeffrey, L. Ament, E. Eliel, van den J. Brink, D. Bouwmeester, Creating and verifying a quantum superposition in a micro-optomechanical system. New J. Phys. **10**(9), 095020 (2008)
16. C.K. Law, H.J. Kimble, Deterministic generation of a bit-stream of single-photon pulses. J. Mod. Opt. **44**(11), 2067–2074 (1997)
17. A.J. Leggett, Macroscopic quantum systems and the quantum theory of measurement. Prog. Theor. Phys. Suppl. **69**, 80 (1980)
18. S. Mancini, D. Vitali, P. Tombesi, Scheme for teleportation of quantum states onto a mechanical resonator. Phys. Rev. Lett. **90**, 137901 (2003)
19. P. Marek, R. Filip, Noise resilient quantum interface based on QND interaction. arXiv:1002.0225
20. F. Marquardt, S.M. Girvin, Optomechanics (a brief review) . Physics **2**, 40 (2009)

21. F. Marquardt, J.P. Chen, A.A. Clerk, S.M. Girvin, Quantum theory of cavity-assisted sideband cooling of mechanical motion. Phys. Rev. Lett. **99**, 093902 (2007)
22. I. Martin, W.H. Zurek, Measurement of energy eigenstates by a slow detector. Phys. Rev. Lett. **98**, 120401 (2007)
23. J. McKeever, A. Boca, A.D. Boozer, R. Miller, J.R. Buck, A. Kuzmich, H.J. Kimble, Deterministic generation of single photons from one atom trapped in a cavity. Science **303**, 1992–1994 (2004)
24. H. Miao, S. Danilishin, T. Corbitt, Y. Chen, Standard quantum limit for probing mechanical energy quantization. Phys. Rev. Lett. **103**, 100402 (2009)
25. H. Miao, S. Danilishin, H. Mueller-Ebhardt, H. Rehbein, K. Somiya, Y. Chen, Probing macroscopic quantum states with a sub-Heisenberg accuracy. Phys. Rev. A **81**, 012114 (2010)
26. G.J. Milburn, Classical and quantum conditional statistical dynamics. Quantum Semiclassical Opt.: J. Eur. Opt. Soc. Part B **8**(1), 269 (1996)
27. H. Mueller-Ebhardt, H. Rehbein, C. Li, Y. Mino, K. Somiya, R. Schnabel, K. Danzmann, Y. Chen, Quantum-state preparation and macroscopic entanglement in gravitational-wave detectors. Phys. Rev. A **80**, 043802 (2009)
28. A.D. O'Connell, M. Hofheinz, M. Ansmann, R.C. Bialczak, M. Lenander, E. Lucero, M. Neeley, D. Sank, H. Wang, M. Weides, J. Wenner, J.M. Martinis, A.N. Cleland, Quantum ground state and single-phonon control of a mechanical resonator. Nature **464**, 697–703 (2010)
29. O. Romero-Isart, M. L. Juan, R. Quidant, J. I. Cirac, Toward quantum superposition of living organisms. arXiv:0909.1469, 2009
30. J.D. Thompson, B.M. Zwickl, A.M. Jayich, F. Marquardt, S.M. Girvin, J.G.E. Harris, Strong dispersive coupling of a high-finesse cavity to a micromechanical membrane. Nature **452**, 72–75 (2008)
31. I. Wilson-Rae, N. Nooshi, W. Zwerger, T.J. Kippenberg, Theory of ground state cooling of a mechanical oscillator using dynamical backaction. Phys. Rev. Lett. **99**, 093901 (2007)
32. J. Zhang, K. Peng, S.L. Braunstein, Quantum-state transfer from light to macroscopic oscillators. Phys. Rev. A **68**, 013808 (2003)

Chapter 11
Probing Macroscopic Quantum States

11.1 Preface

In this chapter, we consider a subsequent verification stage which probes the prepared macroscopic quantum state, and verifies the quantum dynamics. By adopting an optimal time-dependent homodyne detection method, in which the phase of the local oscillator varies in time, the conditional quantum state can be characterized below the Heisenberg limit, thereby achieving a quantum tomography. In the limiting case of no readout loss, such a scheme evades measurement-induced back-action, which is identical to the variational-type measurement scheme invented by Vyatchanin et al., but in the context for detecting gravitational waves (GWs). To motivate Macroscopic Quantum Mechanics (MQM) experiments with future GW detectors, we mostly focus on the parameter regime where the characteristic measurement frequency is much higher than the oscillator frequency and the classical noises are Markovian, which captures the main features of a broadband GW detector. In addition, we discuss verifications of Einstein-Podolsky-Rosen-type entanglement between macroscopic test masses in future GW detectors, which enables us to test one particular version of gravity decoherence conjectured by Diósi and Penrose. This is a joint research effort by Stefan Danilishin, Helge Müller-Ebhardt, Henning Rehbein, Kentaro Somiya, Yanbei Chen, and myself. It is published in Phys. Rev. A **81**, 012114 (2010).

11.2 Introduction

Due to recent significant advancements in fabricating low-loss optical, electrical and mechanical devices, we will soon be able to probe behaviors of macroscopic mechanical oscillators in the quantum regime. This will not only shed light on quantum-limited measurements of various physical quantities, such as a weak force, but also help us to achieve a better understanding of quantum mechanics on macroscopic scales.

As a premise of investigating macroscopic quantum mechanics (MQM), the mechanical oscillator should be prepared close to being in a pure quantum state. To achieve this, there are mainly three approaches raised in the literature: (i) The first and the most transparent approach is to cool down the oscillator by coupling it to an additional heat bath that has a temperature T_{add} much lower than that of the environment T_0. As a result, the oscillator will achieve an effective temperature given by $T_{\text{eff}} = (T_0 \gamma_m + T_{\text{add}} \Gamma_{\text{add}})/(\gamma_m + \Gamma_{\text{add}})$, with γ_m and Γ_{add} denoting the damping due to coupling to the environment and the additional heat bath, respectively. In the strong-damping regime, with $\Gamma_{\text{add}} \gg \gamma_m$, we achieve the desired outcome with $T_{\text{eff}} \approx T_{\text{add}}$. Since the typical optical frequency ω_0 can be much higher than $k_B T_0/\hbar$, a coherent optical field can be effectively behaved as a zero-temperature heat bath. Indeed, by coupling an oscillator parametrically to an optical cavity, many state-of-the-art experiments have demonstrated significant cooling of the oscillator, achieving a very low thermal occupation number [2, 13, 14, 19, 23, 29, 33, 38, 42, 50, 51, 52, 54, 55]. A similar mechanism also applies to electromechanical systems as demonstrated in the experiments [4, 24, 25, 45]; (ii) The second approach is to introduce additional damping via feedback, i.e., the so-called cold-damping. The feedback loop modifies the dynamics of the oscillator in a way similar to the previous cooling case. Such an approach has also been realized experimentally [11, 12, 48]. If the intrinsic mechanical and electrical/optical qualities of the coupled system are high, those cooling and cold-damping experiments can eventually achieve the quantum ground state of a mechanical oscillator [15, 17, 22, 36, 37, 49, 57, 60, 62]; (iii) The third approach is to construct a conditional quantum state of the mechanical oscillator via continuous position measurements. Quantum mechanically, if the oscillator position is being continuously monitored, a certain classical trajectory in the phase space can be mapped out, and the oscillator is projected into *a posteriori state* [3], which is also called a conditional quantum state [18, 21, 26, 40, 43, 44]. Given an ideal continuous measurement without loss, the resulting conditional quantum state of the oscillator is a pure state.

Recently, we theoretically investigated this third approach for general linear position measurements, in great detail [44]. The analysis of this work is independent of the scale and mass of the oscillator—these parameters will only modify the structure of resulting noises. In particular, we applied our formalism to discuss MQM experiments with macroscopic test masses in future gravitational-wave (GW) detectors. We demonstrated explicitly that, given the noise budget for the design sensitivity, next-generation GW detectors such as Advanced LIGO [27] and Cryogenic Laser Interferometer Observatory (CLIO) [41] can prepare nearly pure Gaussian quantum states and create Einstein-Podolsky-Rosen (EPR) type entanglement between macroscopic test masses. Besides, we showed that the free-mass *Standard Quantum Limit* (SQL) [8, 9] for the detection sensitivity:

$$S_x^{\text{SQL}}(\Omega) = \frac{2\hbar}{m \Omega^2}, \quad (11.1)$$

where m is mass of the probing test mass, and Ω is the detection frequency. This limit also serves as a benchmark for MQM experiments with GW detectors.

11.2 Introduction

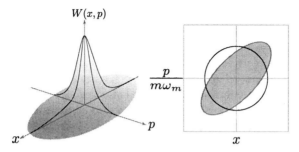

Fig. 11.1 (*color online*)A schematic plot of a Wigner function W(x,p) (*left*) and the corresponding uncertainty ellipse for the covariance matrix $\mathbf{V}^{\text{cond}}$ (this can be viewed as a projection of the Wigner function). The *center* of the plot is given by the conditional mean (x^{cond}, p^{cond}). The Heisenberg limit is shown as a unit circle with radius given by the zero-point fluctuation $\hbar/(2m\omega_m)$. For a pure Gaussian conditional quantum state, the area of the ellipse, i.e., $\pi \det \mathbf{V}^{\text{cond}}/(2m\omega_m)^2$, is also equal to that of the Heisenberg limit. Therefore, the uncertainty product $\det \mathbf{V}^{\text{cond}}$ can be used as an appropriate figure of merit for quantifying the purity of a quantum state. From Ref. [39]

More concretely, a Gaussian conditional quantum state is fully described by its Wigner function, as shown schematically in Fig. 11.1. This is given by:

$$W(x, p) = \frac{1}{2\pi \sqrt{\det \mathbf{V}^{\text{cond}}}} \exp\left[-\frac{1}{2}\vec{X}\,\mathbf{V}^{\text{cond}-1}\,\vec{X}^T\right]. \quad (11.2)$$

Here, $\vec{X} = [x - x^{\text{cond}}, p - p^{\text{cond}}]$ with x^{cond} and p^{cond} denoting the conditional means of oscillator position x and momentum p, and $\mathbf{V}^{\text{cond}}$ is the covariance matrix between position and momentum. The purity of the conditional quantum state can be quantified by the uncertainty product, which is defined as:

$$U \equiv \frac{2}{\hbar}\sqrt{\det \mathbf{V}^{\text{cond}}} = \frac{2}{\hbar}\sqrt{V_{xx}^{\text{cond}} V_{pp}^{\text{cond}} - V_{xp}^{\text{cond}2}}, \quad (11.3)$$

which is also proportional to the square root of the area of the uncertainty ellipse as shown in Fig. 11.1. In Ref. [44], we related this uncertainty product U of the conditional quautum state of test masses in GW detectors to the SQL-beating ratio of the classical noise, and also the amount of entanglement between test masses to the size of the frequency window (i.e., the ratio between upper and lower ends of that frequency window) in which the classical noise goes below the SQL.

A state-preparation stage alone does not provide a complete test of MQM. This is because the measurement data in the state-preparation process only allow us to measure a classical trajectory of the oscillator—quantum fluctuations are only inferred from the noise budget, but are not directly visible. Therefore, the resulting conditional quantum state critically relies on the noise model of the measurement device. If such a noise model is imprecise, it will yield severe discrepancies between the actual quantum state and the conditional one. Therefore, there is a need for a second measurement stage which has to follow up the preparation stage. In this chapter, we will address the above issue by considering a subsequent state-verification procedure, in which we make a tomography of the conditional quantum state obtained during the preparation stage. On the one hand, this verification stage can serve as a check on the specific noise model used to verify the prepared quantum state. On the other hand, if we insert an evolution stage with the oscillator evolving freely before the verification, the quantum dynamics of the oscillator can also be probed, which allows us to study different decoherence effects, and also to check whether a macroscopic mechanical oscillator does evolve in the same way as a quantum harmonic oscillator or not.

Since the conditional quantum state undergoes a random walk in the phase space, as shown schematically in Fig. 11.2, the classical information of the conditional mean, obtained by the preparer from the measurement data, needs to be passed onto the verifier, who will then proceed with a tomography process. Suppose the state preparation stage ends at $t = 0$, and the preparer obtains a conditional quantum state whose Wigner function is $W(x(0), p(0))$. The task of the verifier is to try to reconstruct this Wigner function, by synthesizing marginal distributions of different mechanical quadratures $\hat{X}_\zeta(0)$ from ensemble measurements at $t > 0$, with

$$\hat{X}_\zeta(0) \equiv \hat{x}(0) \cos \zeta + \frac{\hat{p}(0)}{m \omega_m} \sin \zeta, \qquad (11.4)$$

where $\hat{x}(0)$ and $\hat{p}(0)$ denote the oscillator position and momentum at $t = 0$, and ω_m is the oscillation frequency. This process is similar to the optical quantum tomography, where different optical quadratures are measured with homodyne detections [34]. However, there is one significant difference—mechanical quadratures are not directly accessible with linear position measurements, which measure:

$$\hat{x}_q(t) = \hat{x}(0) \cos \omega_m t + \frac{\hat{p}(0)}{m \omega_m} \sin \omega_m t, \qquad (11.5)$$

rather than $\hat{X}_\zeta$. To probe mechanical quadratures, we propose the use of a time-dependent homodyne detection, with the local-oscillator phase varying in time. Given a measurement duration of T_{int}, we can construct an integral estimator, which reads:

$$\hat{X} = \int_0^{T_{\text{int}}} dt\, g(t)\, \hat{x}(t) \propto \hat{x}(0) \cos \zeta' + \frac{\hat{p}(0)}{m \omega_m} \sin \zeta' \qquad (11.6)$$

with $\cos \zeta' \equiv \int_0^{T_{\text{int}}} dt\, g(t) \cos \omega_m t$, and $\sin \zeta' \equiv \int_0^{T_{\text{int}}} dt\, g(t) \sin \omega_m t$. Therefore, a mechanical quadrature $\hat{X}_{\zeta'}$ is probed [cf. Eq. (11.4)]. Here $g(t)$ is some filtering function, which is determined by the time-dependent homodyne phase, and also by the way in which data at different times are combined.

The ability to measure mechanical quadratures does not guarantee the success of a verification process. In order to recover the prepared quantum state, it requires a verification accuracy below the Heisenberg limit. Physically, the output of the verification process is a sum of the mechanical-quadrature signal and some uncorrelated Gaussian noise. Mathematically, it is equivalent to applying a Gaussian filter

11.2 Introduction

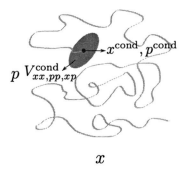

Fig. 11.2 (*color online*) A schematic plot of a random walk of the conditional quantum state, i.e., its Wigner function, in phase space. Its *center* is given by the conditional mean $[x^{\text{cond}}(t), p^{\text{cond}}(t)]$, with the uncertainty given by conditional variances $V_{xx,pp,xp}^{\text{cond}}$. To verify the prepared conditional quantum state, the only knowledge that verifier needs to know is the classical information of the conditional mean provided by the preparer, provided the noises are Markovian. From Ref. [39]

onto the original Wigner function $W(x, p)$ of the prepared state [44], and thus the reconstructed Wigner function is:

$$W_{\text{recon}}(x, p) = \int_{-\infty}^{\infty} dx' dp' \, \psi(x - x', p - p') W(x', p'), \tag{11.7}$$

where the Gaussian filter $\psi(x, p)$ is given by:

$$\psi(x, p) \equiv \frac{1}{2\pi \sqrt{\det \mathbf{V}^{\text{add}}}} \exp\left[-\frac{1}{2} \vec{\xi} \, \mathbf{V}^{\text{add}-1} \vec{\xi}^T \right], \tag{11.8}$$

with $\vec{\xi} = [x, p]$, and $\mathbf{V}^{\text{add}}$ denoting the covariance matrix for the added verification noise. If the prepared quantum state is Gaussian, using the property of Gaussian integration, the reconstructed Wigner function is:

$$W_{\text{recon}}(x, p) = \frac{1}{2\pi \sqrt{\det \mathbf{V}^{\text{recon}}}} \exp\left[-\frac{1}{2} \vec{\xi} \, \mathbf{V}^{\text{recon}-1} \vec{\xi}^T \right], \tag{11.9}$$

and the covariance matrix $\mathbf{V}^{\text{recon}}$ is:

$$\mathbf{V}^{\text{recon}} = \mathbf{V}^{\text{cond}} + \mathbf{V}^{\text{add}}. \tag{11.10}$$

In Fig. 11.3, we show schematically the effects of different levels of verification accuracy given the same prepared conditional quantum state. A sub-Heisenberg accuracy, with an uncertainty area smaller than the Heisenberg limit, is essential for us to obtain a less distorted understanding of the original prepared quantum state. In addition, if the prepared quantum state of the mechanical oscillator is non-Gaussian [6, 28, 30, 31, 35], a sub-Heisenberg accuracy is a necessary condition for unveiling the non-classicality of the quantum state, as shown schematically in Fig. 11.4, and proved rigorously in the Appendix 11.9.1.

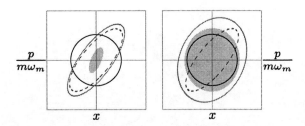

Fig. 11.3 (*color online*) A schematic plot of the uncertainty ellipses of reconstructed states with the same prepared Gaussian quantum state but different levels of verification accuracy, which shows the necessity of a sub-Heisenberg accuracy. The center of the plot is given by the conditional mean (x^{cond}, p^{cond}). The shaded areas correspond to the verification accuracy. The Heisenberg limit is shown by a unit circle. The dashed and solid ellipses represent the prepared state and the reconstructed states respectively. From Ref. [39]

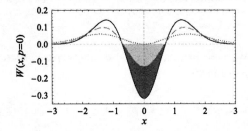

Fig. 11.4 (*color online*) Values of reconstructed Wigner functions on the $p = 0$ plane, i.e., $W_{\mathrm{recon}}(x, p = 0)$, for a single-quantum state, obtained at different levels of verification accuracy. *Solid* curve shows the ideal case with no verification error. *Dashed* and *dotted* curves correspond to the cases with a verification error of 1/4 and 1/2 of the Heisenberg limit, respectively. The *negative* regime (*shaded*), or the non-classicality, vanishes as the verification error increases. This again manifests the importance of a sub-Heisenberg verification accuracy. From Ref. [39]

Verifications of quantum states below the Heisenberg limit also naturally allow us to test whether entanglement between two macroscopic test masses in GW detectors can indeed be established, as predicted in Ref. [43, 44], and how long such an entangled state can survive. Survival of macroscopic entanglement can test one particular version of gravity decoherence conjectured by Diósi [16] and Penrose [47]. For an individual object, it is not entirely clear what is the classical superposition of *pointer* states that gravity decoherence will drive it into. For an entangled state among multiple objects, even though Gaussian, it would naturally have to decay into one that is not entangled, within the gravity decoherence timescale.

As we will show, in order to achieve a sub-Heisenberg accuracy, we need to optimize the local-oscillator phase of the time-dependent homodyne detection as well as the weight with which data collected at different times will be combined. If there is no readout loss, this optimization will automatically give a detection scheme that evades measurement-induced back action, which is the same as the variational-type measurement scheme proposed by Vyatchanin and Matsko [58] for detecting

GW signals with known arrival time. Since, in a single measurement setup, different quadratures do not commute with each other, namely:

$$[\hat{X}_\zeta, \hat{X}_{\zeta'}] = \frac{i\hbar}{m\omega_m} \sin(\zeta - \zeta'), \qquad (11.11)$$

one needs multiple setups, where each makes ensemble measurements of one particular quadrature $\hat{X}_\zeta$ with a sub-Heisenberg accuracy—the synthesis of these measurements yields a quantum tomography.

As a sequel to Ref. [44], and to motivate MQM experiments with future GW detectors, we will also focus on the same parameter regime, where the characteristic measurement frequency is much higher than the oscillator frequency, and the oscillator can be treated as a free mass. In addition, we will consider situations where the spectra of the classical noise can be modeled as being white. Non-Markovianity of noise sources—although they certainly arise in actual GW detectors [44] and will be crucial for the success of a real experiment—is a rather technical issue. This non-Markovianity will not change the results presented here significantly, as we will show and address in a separate paper [in preparation].

This chapter is organized as follows: in Sect. 11.3, we will formulate the system model mathematically by writing down the Heisenberg equations of motion; in Sect. 11.4, we will provide a timeline for a full MQM experiment with preparation, evolution and verification stages, and use simple order-of-magnitude estimates to show that this experimental proposal is indeed plausible; in Sect. 11.6, we will evaluate the verification accuracy in the presence of Markovian noises (largely confirming the order-of-magnitude estimates, but with precise numerical factors); in Sect. 11.7, we will consider verifications of macroscopic quantum entanglement between test masses in GW detectors as a test of gravity decoherence; in Sect. 11.8, we will summarize our main results. In the Appendix, we will present mathematical details for solving the integral equations that we encounter in obtaining the optimal verification scheme.

11.3 Model and Equations of Motion

In this section, we will present a mathematical description of the system model, as shown schematically in the upper left panel of Fig. 11.5 The oscillator position is linearly coupled to coherent optical fields through radiation pressure. Meanwhile, information of the oscillator position flows into the outgoing optical fields continuously. This models a measurement process in an optomechanical system without a cavity, or with a large-bandwidth cavity. The corresponding Heisenberg equations, valid for both preparation and verification stages, are formally identical to classical equations of motion except that all quantities are Heisenberg operators. The oscillator position $\hat{x}$ and momentum $\hat{p}$ satisfy the following equations:

$$\dot{\hat{x}}(t) = \hat{p}(t)/m, \qquad (11.12)$$

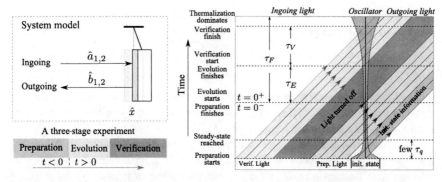

Fig. 11.5 A schematic plot of the system (*upper left* panel) and the corresponding spacetime diagram (*right panel*) showing the timeline of the proposed MQM experiment (see Sect. 11.4.1 for detailed explanations). In this schematic plot, the oscillator position is denoted by $\hat{x}$ which is coupled to the optical fields through radiation pressure. The ingoing and outgoing optical fields are denoted by $\hat{a}_{1,2}$ and $\hat{b}_{1,2}$ with subscripts 1, 2 for the amplitude and phase quadratures, respectively. In the spacetime diagram, the world line of the oscillator is shown by the *middle vertical* line. For clarity, *ingoing* and *outgoing* optical fields are represented by the *left* and *right* regions on opposite sides of the oscillator world line, even though in reality, the optical fields escape from the same side as where they enter. We show light rays during the preparation and verification stages in *red* and *blue*. In between, the *yellow* shaded region describes the evolution stage, with the light turned off for a duration of τ_E. The conditional variance of the oscillator motion is represented by the *shaded* region alongside the *central vertical* line (not drawn to the same scale as for the light propagation). At the beginning of preparation, the conditional variance is dominated by that of the initial state (*orange*). After a transient, it is determined by incoming radiation and measurements. Right after state preparation, we show the expected growth of the conditional variance due to thermal noise alone, and ignoring the effect of back-action noise, which is evaded during the verification process. The verification stage lasts for a duration of τ_V, and it is shorter than τ_F, after which the oscillator will be dominated by thermalization. From Ref. [39]

$$\dot{\hat{p}}(t) = -2\gamma_m \hat{p}(t) - m\,\omega_m^2\,\hat{x}(t) + \alpha\,\hat{a}_1(t) + \hat{\xi}_F(t). \quad (11.13)$$

Here, $\alpha\,\hat{a}_1$ corresponds to the quantum-radiation-pressure noise, or the so-called back-action noise; $\alpha \equiv (\hbar\,m\,\Omega_q^2)^{1/2} = (8\,I_0\,\omega_0\,\hbar/c^2)^{1/2}$ is the coupling constant between the oscillator and the optical fields, with I_0 denoting the optical power, and Ω_q quantifying the characteristic frequency of measurement strength. We have included the fluctuation-dissipation mechanism of the mechanical oscillator by introducing the mechanical damping rate γ_m, and classical-force noise $\hat{\xi}_F$, i.e., the Brownian thermal noise. In the Markovian limit, the correlation function for $\hat{\xi}_F$ is given by[1]:

---

[1] Here $\langle\;\rangle_{\mathrm{sym}}$ stands for a *symmetrized* ensemble average. For a system characterized by a density matrix $\hat{\rho}$, it is defined as

$$\langle \hat{o}_1(t)\,\hat{o}_2(t')\rangle_{\mathrm{sym}} \equiv \mathrm{Tr}\left\{[\hat{o}_1(t)\hat{o}_2(t') + \hat{o}_2(t')\hat{o}_1(t)]\hat{\rho}\right\}/2.$$

11.3 Model and Equation of Motion

$$\langle \hat{\xi}_F(t) \hat{\xi}_F(t') \rangle_{\text{sym}} = S_F^{\text{th}} \delta(t-t')/2, \quad (11.14)$$

where $S_F^{\text{th}} = 4m\gamma_m k_B T_0 \equiv 2\hbar m \Omega_F^2$ and we have defined a characteristic frequency Ω_F for the thermal noise.

The amplitude and phase quadratures of ingoing optical fields $\hat{a}_{1,2}$, and of outgoing optical fields $\hat{b}_{1,2}$, satisfy the following input-output relations:

$$\hat{b}_1(t) = \sqrt{\eta}\,\hat{n}_1(t) + \sqrt{1-\eta}\,\hat{a}_1(t), \quad (11.15)$$

$$\hat{b}_2(t) = \sqrt{\eta}\,\hat{n}_2(t) + \sqrt{1-\eta}\left[\hat{a}_2(t) + \frac{\alpha}{\hbar}\hat{x}(t) + \frac{\alpha}{\hbar}\hat{\xi}_x(t)\right]. \quad (11.16)$$

Here $\hat{n}_{1,2}$, originate from non-unity quantum efficiency of the photodetector for $\eta > 0$. In the paraxial and narrow-band approximation, $\hat{a}_{1,2}$ are related to the electrical-field strength at the central frequency ω_0 by [5, 10, 32]:

$$\hat{E}(t) \equiv \left(\frac{4\pi\hbar\omega_0}{Sc}\right)^{1/2} \{[\bar{a} + \hat{a}_1(t)]\cos\omega_0 t + \hat{a}_2(t)\sin\omega_0 t\} \quad (11.17)$$

with $\bar{a}$ denoting the classical amplitude, and S standing for the effective cross-sectional area of the laser beam. A similar relation also holds for the outgoing fields $\hat{b}_{1,2}$. In addition, they satisfy $[\hat{a}_1(t), \hat{a}_2(t')] = [\hat{b}_1(t), \hat{b}_2(t')] = i\delta(t-t')$. Their correlation functions are:

$$\langle \hat{a}_i(t)\hat{a}_j(t') \rangle_{\text{sym}} = \delta_{ij} e^{\pm 2q} \delta(t-t')/2, \quad (i,j = 1,2) \quad (11.18)$$

where q denotes the squeezing factor ($q = 0$ for a vacuum-state input), with "+" for the amplitude quadrature and "−" for the phase quadrature. Correspondingly, the correlation function for the back-action noise $\alpha\,\hat{a}_1$ is simply

$$\langle \alpha\,\hat{a}_1(t)\,\alpha\,\hat{a}_1(t') \rangle_{\text{sym}} = S_F^{\text{BA}} \delta(t-t')/2, \quad (11.19)$$

with $S_F^{\text{BA}} \equiv e^{2q}\hbar m \Omega_q^2$. In Eq. (11.16), $\hat{\xi}_x$ is the sensing noise. One example is the internal thermal noise, and it is defined as the difference between the center of mass motion and the surface motion of the oscillator which is actually being measured. In the Markovian approximation, it has the following correlation function:

$$\langle \hat{\xi}_x(t) \hat{\xi}_x(t') \rangle_{\text{sym}} = S_x^{\text{th}} \delta(t-t')/2, \quad (11.20)$$

where $S_x^{\text{th}} = \hbar/(m\Omega_x^2)$ and we introduce a characteristic frequency Ω_x for the sensing noise.

Note that the Ω_q, Ω_F, and Ω_x that we have introduced are also the frequencies at which the back-action noise, thermal noise, and sensing noise intersect the SQL [cf. Eq. (11.1)], respectively. They are identical to what were introduced in Ref. [44]. For conveniences of later discussions, we introduce the following dimensionless ratios:

$$\zeta_F = \Omega_F/\Omega_q, \quad \zeta_x = \Omega_q/\Omega_x. \tag{11.21}$$

In addition, we define two characteristic timescales for the measurement and thermal-noise strength as:

$$\tau_q \equiv 1/\Omega_q, \quad \tau_F \equiv 1/\Omega_F. \tag{11.22}$$

11.4 Outline of the Experiment With Order-of-Magnitude Estimate

In this section, we will describe in detail the timeline of a plausible MQM experiment (Sect. 11.4.1), and provide order-of-magnitude estimates of the conditional variance of the prepared quantum state, the evolution of the prepared quantum state, and the verification accuracy in the free-mass regime (Sects. 11.4.2–11.4.4). This will provide qualitatively the requirements on the noise level for the success of an MQM experiment. We will give more rigorous treatments in Sect. 11.6.

11.4.1 Timeline of Proposed Experiment

We have sketched a space-time diagram for the proposed MQM experiment in the right panel of Fig. 11.5—with time going upward, we therefore start from the bottom of the figure.

Lock Acquisition. At the beginning, the mechanical oscillator is in a highly mixed state, and so are the optical fields. Therefore, the first step is to "acquire lock" of the measurement device, and reach a steady-state operation mode, during which several τ_q will have elapsed. From this time onwards, the initial-state information will have been forgotten (propagating outward within the green strip), and the state of the oscillator will be determined by the driving fields, including the classical-force noise and sensing noise, as well as the quantum noise. This will be the start of the state-preparation stage (region above the 45° green strip).

State Preparation. This stage is a steady-state operation of the measurement device. The quantum state of the oscillator is collapsed continuously due to homodyne readouts of the photocurrent. At any instant during state preparation, based on the measured history of the photocurrent (mostly on data within several times τ_q to the past of t), the conditional expectation x^{cond}, p^{cond}) for the oscillator position $\hat{x}$ and momentum $\hat{p}$ can be constructed. The second moments, describable by the covariance matrix between position and momentum, which consists of V_{xx}^{cond}, V_{xp}^{cond} and V_{pp}^{cond}, can be calculated from the noise model of the measurement device—they, together with x^{cond} and p^{cond}, fully determine the quantum state, i.e., the Wigner function of the oscillator at any instant [cf. Eq. (11.2)]. For a Gaussian steady state, the construction of $(x^{\text{cond}}, p^{\text{cond}})$ and the conditional covariance matrix from the history of the photocurrent can be accomplished most easily using Wiener Filtering, as shown in Ref. [44].

The preparation stage terminates at $t = 0$, when $(x^{\text{cond}}, p^{\text{cond}})$ and the covariance matrix will be determined by data from several $-\tau_q$ up to 0 as shown by the red strip.

State Evolution. If we want to investigate the quantum dynamics of the oscillator and study various decoherence effects, we can delay the verification process and allow the oscillator to freely evolve with the interaction light turned off (as represented by the yellow strip). During this period, the thermal noise will induce diffusions of the oscillator position and momentum, thus increasing the conditional variance as shown schematically by the broadening of the shaded region alongside the oscillator world line. If there were any additional decoherence effect, the variance would grow faster than the case with the thermal decoherence alone. A follow-up verification allows us to check whether additional decoherence mechanisms, such as the gravity decoherence conjectured by Diósi [16] and Penrose [47], exist or not.

State Verification. After the evolution stage, the verification stage starts (represented by blue strip). We intentionally use different colors to label the preparation light and verification light—symbolizing the fact that, in principle, a different observer (verifier) could perform the verification process, and verify the quantum state by him/herself. The only knowledge from the preparer would be the conditional expectation x^{cond} and p^{cond}, if all noise sources are Markovian. The verifier uses a time-dependent homodyne detection and collects the data from measuring the photocurrents. The verification process lasts for a timescale of τ_V between the characteristic measurement timescale τ_q and the thermal decoherence timescale τ_F, after which diffusions of $\hat{x}$ and $\hat{p}$ in the phase space become much larger than the Heisenberg limit. Based upon the measurement data, the verifier can construct an integral estimator for one particular mechanical quadrature [cf. Eq. 11.6].

The above three stages have to be repeated many times before enough data are collected to build up reliable statistics. After finishing the experiment, the verifier will obtain a reconstructed quantum state of the mechanical oscillator, and then can proceed to compare with the preparer, and to interpret the results.

11.4.2 Order-of-Magnitude Estimate of the Conditional Variance

In this and the following two subsections, we will provide order-of-magnitude estimates for a three-staged MQM experiment including preparation, evolution and verification stages. This gives us physical insight into the different timescales involved in an MQM experiment, and also into the qualitative requirements for an experimental realization. We will justify those estimates based upon more careful treatments in the next several sections.

Based upon the measurement data from several $-\tau_q$ to 0, one can construct a conditional quantum state for the mechanical oscillator. Suppose that the phase quadrature of the outgoing fields is being measured and the photodetection is ideal with $\eta = 0$.

Given a measurement timescale of τ (measuring from $-\tau$ to 0), variances for the oscillator position and momentum at $t = 0$, in the free-mass regime with $\omega_m \to 0$, are approximately equal to [cf. Eqs. (11.12), (11.13), (11.15) and (11.16)]

$$\delta x^2(0) \sim S_x^{\text{tot}}/\tau + \tau^3 S_F^{\text{tot}}/m^2 \sim N_x^{\frac{3}{4}} N_F^{\frac{1}{4}} \delta x_q^2, \qquad (11.23)$$

$$\delta p^2(0) \sim m^2 S_x^{\text{tot}}/\tau^3 + \tau S_F^{\text{tot}} \sim N_x^{\frac{1}{4}} N_F^{\frac{3}{4}} \delta p_q^2. \qquad (11.24)$$

Here, $S_F^{\text{tot}} \equiv S_F^{\text{BA}} + S_F^{\text{th}}$ [cf. Eqs. (11.14) and (11.19)] and $S_x^{\text{tot}} \equiv S_x^{\text{sh}} + S_x^{\text{th}}$ with S_x^{sh} denoting the shot noise due to $\hat{a}_2$ [cf. Eqs. (11.16) and (11.20)]; we have defined

$$N_x \equiv 1 + 2\zeta_x^2, \quad N_F \equiv 1 + 2\zeta_F^2, \qquad (11.25)$$

while

$$\delta x_q^2 \equiv \hbar/(2m\Omega_q), \quad \delta p_q^2 \equiv \hbar m \Omega_q/2. \qquad (11.26)$$

The optimal measurement timescale is given by $\tau \sim \tau_q$. The purity of the prepared conditional quantum state at $t = 0$ is approximately equal to [cf. Eq. (11.3)]

$$U(0) \sim \frac{2}{\hbar} \delta x(0) \, \delta p(0) \sim N_x N_F. \qquad (11.27)$$

If the classical noises are low, namely, $N_x \sim N_F \sim 1$, the conditional quantum state will be pure, with $U(0) \sim 1$. For future GW detectors such as AdvLIGO, both ζ_x and ζ_F will be around 0.1, and such a low classical-noise budget clearly allows us to prepare nearly pure quantum states of the macroscopic test masses.

11.4.3 Order-of-Magnitude Estimate of State Evolution

During the evolution stage, the uncertainty ellipse of the conditional quantum state will rotate at the mechanical frequency in the phase space, and meanwhile there is a growth in the uncertainty due to thermal decoherence as shown schematically in Fig. 11.6. Given a strong measurement, the variance of the resulting conditional quantum state in position $\delta x^2(0)$ will be approximately equal to δx_q^2 as shown explicitly in Eq. (11.23) with $N_x, N_F \sim 1$. It is much smaller than the zero-point uncertainty of an ω_m oscillator, which is given by $\hbar/(2m\omega_m)$. Therefore, the conditional quantum state of the oscillator is highly squeezed in position. The position uncertainty contributed by the initial-momentum will be comparable to that of the initial-position uncertainty after a evolution duration of τ_q. This can be directly seen from an order-of-magnitude estimate. In the free-mass regime,

$$\hat{x}(t) \sim \hat{x}(0) + \frac{\hat{p}(0)}{m} t. \qquad (11.28)$$

11.4 Outline of the Experiment With Order-of-Magnitude Estimate

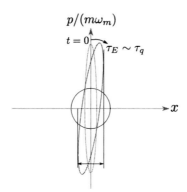

Fig. 11.6 (*color online*) Rotation and diffusion of a highly position-squeezed conditional quantum state, prepared by a strong measurement with $\Omega_q \gg \omega_m$. The initial-momentum uncertainty will contribute an uncertainty in the position comparable to the initial-position uncertainty, when the evolution duration $\tau_E \sim \tau_q$. From Ref. [39]

For an evolution duration of τ_E, the corresponding variance in position is:

$$\delta x^2(\tau_E) \sim \delta x^2(0) + \frac{\delta p^2(0)}{m^2}\tau_E^2 \sim \delta x^2(0)[1 + (\Omega_q \tau_E)^2]. \quad (11.29)$$

The contribution from the initial-momentum uncertainty (the second term) will become important when $\Omega_q \tau_E \sim 1$, or equivalently $\tau_E \sim \tau_q$.

Apart from a rotation, the uncertainty ellipse will also grow due to thermal decoherence. Variances in the position and momentum contributed by thermal decoherence are approximately given by [cf. Eqs. (11.12) and (11.13)]

$$\delta x_{\text{th}}^2(\tau_E) \sim \tau_E^3 S_F^{\text{th}}/m^2 = \zeta_F^2(\Omega_q \tau_E)^3 \delta x_q^2, \quad (11.30)$$

$$\delta p_{\text{th}}^2(\tau_E) \sim \tau_E S_F^{\text{th}} = \zeta_F^2(\Omega_q \tau_E)\delta p_q^2. \quad (11.31)$$

The growth in the uncertainty ellipse will simply be:

$$U^{\text{th}}(\tau_E) \sim \frac{2}{\hbar}\delta x_{\text{th}}(\tau_E)\delta p_{\text{th}}(\tau_E) \sim \zeta_F^2(\Omega_q \tau_E)^2 = (\tau_E/\tau_F)^2. \quad (11.32)$$

When $\tau_E > \tau_F$, $U^{\text{th}}(\tau_E) > 1$, and the conditional quantum state will be dominated by thermalization.

If there were any additional decoherence effect, the growth in the uncertainty would be much larger than what has been estimated here. A subsequent verification stage can serve as a check.

11.4.4 Order-of-Magnitude Estimate of the Verification Accuracy

To verify the prepared conditional quantum state, the oscillator position needs to be measured for a finite duration to obtain information about $\hat{x}(0)$ and $\hat{p}(0)$ [cf. Eqs. (11.15) and (11.16)] or about $\hat{x}(\tau_E)$ and $\hat{p}(\tau_E)$ if the evolution stage is inserted. In order for an entire state characterization to be possible, one might then expect that an oscillation period must pass, and during this period, the thermal noise should cause an insignificant diffusion of the oscillator momentum compared with its zero-point uncertainty, which requires [44]:

$$\frac{k_B T_0}{\hbar \omega_m} < Q_m \tag{11.33}$$

with $Q_m \equiv \omega_m/(2\gamma_m)$ denoting the mechanical quality factor. This requirement is unnecessary if the initial quantum state is prepared by a strong measurement. As we have mentioned in the previous subsection, the resulting conditional quantum state is highly squeezed in position, and the initial-momentum uncertainty will make a significant contribution to the uncertainty in position after $\tau > \tau_q$. This means, depending on the particular strategy, one can extract $\hat{x}$ and $\hat{p}$ below the levels of δx_q and δp_q, respectively, as long as one is able to measure oscillator position with an accuracy better than δx_q, within a timescale of several τ_q. This is certainly possible if the measurement-induced back-action is evaded.

To evade the measurement-induced back-action, one notices the fact that the amplitude quadrature $\hat{b}_1$ contains $\hat{a}_1$, which is responsible for the back action, and meanwhile the phase quadrature $\hat{b}_2$ contains the information of oscillator position, part of which is contributed by the back action [cf. Eqs. (11.12)–(11.16)]. Therefore, if we measure particular combinations of $\hat{b}_1$ and $\hat{b}_2$ at different times, by summing up those measurements, we will be able to cancel the back action and obtain a back-action-evading (BAE) estimator for a given mechanical quadrature. Such a cancelation mechanism is only limited by the readout loss ($\eta \neq 0$), which introduces uncorrelated vacuum fluctuations.

We can make an order-of-magnitude estimate to show that a sub-Heisenberg accuracy can be indeed achieved. With the BAE technique, the force noise that limits the verification accuracy will only contain the thermal-noise part. Similar to Eqs. (11.23) and (11.24) but with S_F^{tot} replaced by S_F^{th}, the variances in position and momentum during the verification stage are simply:

$$\delta x_V^2 \sim S_x^{\text{tot}}/\tau + \tau^3 S_F^{\text{th}}/m^2 \sim N_x^{3/4} \zeta_F^{1/2} \delta x_q^2, \tag{11.34}$$

$$\delta p_V^2 \sim m^2 S_x^{\text{tot}}/\tau^3 + \tau S_F^{\text{th}} \sim N_x^{1/4} \zeta_F^{3/2} \delta p_q^2. \tag{11.35}$$

Here the optimal verification timescale would be $\tau_V \sim \zeta_F^{-1/2} \tau_q$, and $\tau_q < \tau_V < \tau_F$. A summarizing figure of merit for the verification accuracy is approximately given by:

11.4 Outline of the Experiment With Order-of-Magnitude Estimate

$$U^{\text{add}}|_{\text{BAE}} \sim \frac{2}{\hbar}\delta x_V \delta p_V \sim N_x^{1/2}\zeta_F. \tag{11.36}$$

A sub-Heisenberg accuracy can be achieved when $\zeta_F < 1$. Note that this error can be made arbitrarily small by lowering ζ_F indefinitely, i.e., a very strong measurement. If phase-squeezed light is injected during the verification stage, we would have

$$U^{\text{add}}|_{\text{BAE}} \sim (e^{-2q} + 2\zeta_x^2)^{1/2}\zeta_F = \sqrt{\frac{\Omega_F^2}{\Omega_q^2 e^{2q}} + \frac{2\Omega_F^2}{\Omega_x^2}}. \tag{11.37}$$

Increasing the squeezing factor always improves our verification sensitivity, with a limit of

$$U^{\text{add}}_{\text{lim}}|_{\text{BAE}} \sim \Omega_F/\Omega_x = \zeta_x\zeta_F, \tag{11.38}$$

which can be much lower than unity in the case of future GW detectors, or of any low-noise measurement device.

Had we not evaded the back-action noise, we would have $\sqrt{N_F}$ in the place of ζ_F, which means $\delta x_V \delta p_V$ would be Heisenberg-limited—unless different squeezing factors are assumed. For low squeezing (i.e., $e^{\pm 2q}$ larger than both ζ_x and ζ_F), we need phase-squeezing for $\hat{x}$ observation, and amplitude squeezing for $\hat{p}$ observation, with

$$U^{\text{add}}|_{\text{without BAE}} \sim e^{-q}, \tag{11.39}$$

which is a significant factor $(1/\zeta_F)$ worse than in the BAE scheme. Even though there exists an optimal squeezing factor that this scheme can apply, which is:

$$U^{\text{add}}_{\text{opt}}|_{\text{without BAE}} \sim \zeta_x, \tag{11.40}$$

it is still worse than the limiting situation of the BAE scheme [cf. Eq. (11.38)] by a factor of $1/\zeta_F$ ($\gg 1$).

11.5 The Conditional Quantum State and its Evolution

The previous order-of-magnitude estimates provide us with a qualitative picture of an MQM experiment, especially in the free-mass regime where future GW detectors will be operating. As long as ζ_F and ζ_x are smaller than unity, i.e., the classical noise goes below the SQL around the most sensitive frequency band ($\Omega \sim \Omega_q$) of the measurement device, not only can we prepare a nearly pure quantum state, but also we can make a sub-Heisenberg tomography of the prepared state. In this and following sections, we will provide more rigorous treatments by directly analyzing the detailed dynamics of the system.

11.5.1 The Conditional Quantum State Obtained From Wiener Filtering

The rigorous mathematical treatment of state preparation has been given in Ref. [44]. The main idea is to treat the conditional quantum state preparation as a classical filtering problem, which is justified by the fact that the outgoing optical quadratures $\hat{b}_{1,2}$ at different times commute with each other, the same as in a classical random process. For such a Gaussian linear system, the Wiener filter, satisfying the minimum mean-square error criterion, allows us to obtain an optimal estimate for the quantum state of the oscillator, i.e., the conditional quantum state. Based upon the measurement data $y(t)$ ($t < 0$), conditional means for the oscillator position and momentum at $t = 0$ can be constructed as [cf. Eq. (14) of Ref. [44]]

$$x^{\text{cond}}(0) \equiv \langle \hat{x}(0) \rangle^{\text{cond}} = \int_{-\infty}^{0} dt\, K_x(-t) y(t), \qquad (11.41)$$

$$p^{\text{cond}}(0) \equiv \langle \hat{p}(0) \rangle^{\text{cond}} = \int_{-\infty}^{0} dt\, K_p(-t) y(t). \qquad (11.42)$$

Here K_x and K_p are causal Wiener filters. The covariance matrix is given by [cf. Eq. (15) of Ref. [44]]:

$$\mathbf{V}^{\text{cond}}_{o_i o_j}(0) = \langle \hat{o}_i(0) \hat{o}_j(0) \rangle^{\text{cond}}_{\text{sym}} - \langle \hat{o}_i(0) \rangle^{\text{cond}} \langle \hat{o}_j(0) \rangle^{\text{cond}}, \qquad (11.43)$$

where $i, j = 1, 2$ and $\hat{o}_1, \hat{o}_2$ denote $\hat{x}, \hat{p}$, respectively. In the free-mass regime, we showed that [cf. Eqs. (52)–(54) in Ref. [37]]:

$$\mathbf{V}^{\text{cond}}(0) = \begin{bmatrix} N_F^{\frac{1}{4}} N_x^{\frac{3}{4}} \sqrt{2}\delta x_q^2 & N_F^{\frac{1}{2}} N_x^{\frac{1}{2}} \hbar/2 \\ N_F^{\frac{1}{2}} N_x^{\frac{1}{2}} \hbar/2 & N_F^{\frac{3}{4}} N_x^{\frac{1}{4}} \sqrt{2}\delta p_q^2 \end{bmatrix}. \qquad (11.44)$$

With conditional means and variances, the Wigner function or equivalently the conditional quantum state is uniquely defined [cf. Eq. (11.2)]. Correspondingly, purity of the conditional quantum state is quantified by

$$U(0) = \frac{2}{\hbar}\sqrt{\det \mathbf{V}^{\text{cond}}(0)} = N_x N_F. \qquad (11.45)$$

This simply justifies the order-of-magnitude result presented in Eq. (11.27).

11.5.2 Evolution of the Conditional Quantum State

In the following discussions, we will analyze how such a conditional quantum state evolves during the evolution stage. On the one hand, this confirms the qualitative

11.5 The Conditional Quantum State and its Evolution

results presented in Sect. 11.4.3. On the other hand, it provides a quantitative understanding of the timescale for the later verification stage.

The equations of motion for the oscillator during the evolution stage are given by Eqs. (11.12) and (11.13) except that there is no radiation pressure, as the light is turned off.[2] For simplicity, and also in consideration of the case in a realistic experiment, we will assume an oscillator with a high quality factor, i.e., $\omega_m \gg \gamma_m$. Within a timescale much shorter than $1/\gamma_m$, the oscillator can be well-approximated as a free oscillator. Correspondingly, the analytical solution for the oscillator position reads:

$$\hat{x}(t) = \hat{x}_q(t) + \int_0^\infty dt' \, G_x(t-t') \hat{\xi}_F(t'). \tag{11.46}$$

Here the free quantum oscillation $\hat{x}_q(t)$ of the oscillator is given by Eq. (11.5). We have defined the Green's function as:

$$G_x(t) = \Theta(t) \frac{\sin(\omega_m t)}{m \omega_m}, \tag{11.47}$$

with $\Theta(t)$ denoting the Heaviside function.

Given an evolution duration of τ_E, from Eqs. (11.14) and (11.46) the corresponding covariance matrix evolves as

$$\mathbf{V}(\tau_E) = \mathbf{R}_\Phi^T \mathbf{V}^{\text{cond}}(0) \mathbf{R}_\Phi + \frac{S_F^{\text{th}}}{8 m^2 \omega_m^3} \begin{bmatrix} 2\Phi - \sin 2\Phi & 2m\omega_m \sin^2 \Phi \\ 2m\omega_m \sin^2 \Phi & m^2 \omega_m^2 (2\Phi + \sin 2\Phi) \end{bmatrix}, \tag{11.48}$$

where $\Phi \equiv \omega_m \tau_E$, and the rotation matrix $\mathbf{R}_\Phi$ is given by:

$$\mathbf{R}_\Phi = \begin{bmatrix} \cos \Phi & -m\omega_m \sin \Phi \\ (m\omega_m)^{-1} \sin \Phi & \cos \Phi \end{bmatrix}. \tag{11.49}$$

The first term in Eq. (11.48) represents a rotation of the covariance matrix $\mathbf{V}^{\text{cond}}(0)$ due to the free quantum oscillation of the oscillator; the second term is contributed by thermal decoherence which causes an increase in the uncertainty.

In the free-mass regime and the case of $\omega_m \tau_E \ll 1$, elements of the covariance matrix can be expanded as series of Φ. Up to the leading order in Φ, we obtain

$$V_{xx}(\tau_E) = V_{xx}^{\text{cond}} + \frac{4\delta x_q^2}{\hbar} V_{xp}^{\text{cond}} \Omega_q \tau_E + \frac{\delta x_q^2}{\delta p_q^2} V_{pp}^{\text{cond}} (\Omega_q \tau_E)^2 + 2\delta x_q^2 \zeta_F^2 \frac{(\Omega_q \tau_E)^3}{3}, \tag{11.50}$$

---

[2] Were the light turned on, the back action could still be evaded as long as one measures the amplitude quadrature $\hat{a}_1$ during this period and take them into account during data processing. Since no information of the oscillator position (contained in the phase quadrature of the outgoing light) is collected, this is equivalent to the case with the light turned off.

$$V_{xp}(\tau_E) = V_{xp}^{\text{cond}} + \frac{\hbar}{2\delta p_q^2} V_{pp}^{\text{cond}} \Omega_q \tau_E + \frac{\hbar}{2} \zeta_F^2 (\Omega_q \tau_E)^2, \qquad (11.51)$$

$$V_{pp}(\tau_E) = V_{pp}^{\text{cond}} + 2\delta p_q^2 \zeta_F^2 \Omega_q \tau_E, \qquad (11.52)$$

with $V_{xx,xp,pp}^{\text{cond}}$ denoting the elements of $\mathbf{V}^{\text{cond}}(0)$. Up to the leading order in $\Omega_q \tau_E$, the uncertainty product of the resulting quantum state is:

$$U(\tau_E) = \frac{2}{\hbar} \sqrt{\det \mathbf{V}(\tau_E)} \approx U(0) + \frac{V_{xx}^{\text{cond}}}{\delta x_q^2} (\tau_E/\tau_F)^2, \qquad (11.53)$$

with τ_F defined in Eq. (11.2). The second term is contributed by the thermal decoherence and can be viewed as $U^{\text{th}}(\tau_E)$. Those formulas recover the results in Eqs. (11.29)–(11.32), but now with precise numerical factors. As we can conclude from Eq. (11.53), in order for a sub-Heisenberg tomography to be possible, the later verification stage should finish within a timescale of τ_F, after which the contribution from the thermal noise gives $U^{\text{th}}(\tau_F) \sim 1$.

11.6 State Verification in the Presence of Markovian Noises

In this section, we will treat the follow-up state verification stage with Markovian noises in detail. This can justify the order-of-magnitude estimate we have derived in Sect. 11.4.4. In addition, we will show explicitly how to construct the optimal verification scheme that gives a sub-Heisenberg accuracy.

11.6.1 A Time-Dependent Homodyne Detection and Back-Action-Evasion

In this subsection, we will analyze the time-dependent homodyne detection which enables us to probe mechanical quadratures. We will further show how the BAE scheme can be constructed. The BAE scheme is optimal only when there is no readout loss ($\eta = 0$). We will consider more general situations, and derive the corresponding optimal verification scheme in the next subsection.

The equations of motion for the oscillator during the verification stage ($t > \tau_E$) are given by Eqs. (11.12) and (11.13). The corresponding solution for the oscillator position is different from Eq. (11.46) due to the presence of the back-action noise which starts to act on the oscillator at $t = \tau_E$. Specifically, it reads

$$\hat{x}(t) = \hat{x}_q(t) + \int_{\tau_E}^{\infty} dt' \, G_x(t-t') [\alpha \, \hat{a}_1(t') + \hat{\xi}_F(t')]. \qquad (11.54)$$

11.6 State Verification in the Presence of Markovian Noises

Here the free quantum oscillation $x_q(t)$ is the signal that we seek to probe during the verification stage. For optical quadratures, the equations of motion are given by Eqs. (11.15) and (11.16). From these equations, we notice that among the outgoing fields: $\hat{b}_1$ is pure noise, while $\hat{b}_2$ contains both signal $\hat{x}_q(t)$ and noise. In order to highlight this, we rewrite $\hat{b}_{1,2}$ as:

$$\hat{b}_1(t) = \sqrt{1-\eta}\,\hat{n}_1(t) + \sqrt{\eta}\,\hat{a}_1(t) \equiv \delta\hat{b}_1(t), \tag{11.55}$$

$$\hat{b}_2(t) = \delta\hat{b}_2(t) + \sqrt{1-\eta}\,(\alpha/\hbar)\,\hat{x}_q(t), \tag{11.56}$$

with [cf. Eq. (11.54)]:

$$\delta\hat{b}_2(t) \equiv \sqrt{\eta}\,\hat{n}_2(t) + \sqrt{1-\eta}\left\{\hat{a}_2(t) + \frac{\alpha}{\hbar}\hat{\xi}_x(t) \right.$$
$$\left. + \frac{\alpha}{\hbar}\int_{\tau_E}^{\infty} dt'\, G_x(t-t')[\alpha\,\hat{a}_1(t') + \hat{\xi}_F(t')]\right\}. \tag{11.57}$$

In this way, we can directly see that $\hat{a}_1$ which causes the back-action is contained in both the amplitude quadrature $\hat{b}_1$ and the phase quadrature $\hat{b}_2$. Therefore, by measuring an appropriate combination of the two output quadratures, we will be able to remove the effects of the back-action noise that is imposed on the oscillator during the verification process at $t > \tau_E$. Searching for such an optimal combination is the main issue to be addressed in this section.

As mentioned in the introduction part, to probe mechanical quadratures and their distributions, a time-dependent homodyne detection needs to be applied [cf. Eq. (11.6)]. Specifically, the outgoing optical field:

$$\hat{B}_{\text{out}}(t) = \hat{b}_1(t)\cos\omega_0 t + \hat{b}_2(t)\sin\omega_0 t \tag{11.58}$$

at $t > \tau_E$ is mixed with a strong local-oscillator light $L(t)$ whose phase angle ϕ_{os} is time-dependent, as shown schematically in Fig. 11.7, namely:

$$L(t) = L_0 \cos[\omega_0 t - \phi_{os}(t)] \tag{11.59}$$

with L_0 a time-independent constant. Through a low-pass filtering (with a bandwidth much smaller than ω_0) of the beating signal, the resulting photocurrent is:

$$\hat{i}(t) \propto \overline{2\hat{B}_{\text{out}}(t)L(t)}$$
$$= L_0\,\hat{b}_1(t)\cos\phi_{os}(t) + L_0\,\hat{b}_2(t)\sin\phi_{os}(t), \tag{11.60}$$

where the overline means averaging over many optical-oscillation periods. Note that the Heisenberg operators for the photocurrent at different times commute with each other, i.e.,

$$[\hat{i}(t), \hat{i}(t')] = 0, \tag{11.61}$$

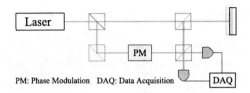

Fig. 11.7 (*color online*) A schematic plot of time-dependent homodyne detection. The phase modulation of the local oscillator light varies in time. From Ref. [39]

and are therefore *simultaneously measurable*, as expected. Based on the measurement results of $\hat{i}(t)$ from τ_E to T_{int}, we can construct the following weighted quantity $\hat{Y}$, with a weight function $W(t)$:

$$\hat{Y} = \int_0^{T_{\text{int}}} \Theta(t - \tau_E) W(t) \hat{i}(t) dt \equiv (g_1 | \hat{b}_1) + (g_2 | \hat{b}_2). \tag{11.62}$$

Here, the Heaviside function $\Theta(t - \tau_E)$ manifests the fact that the verification stage starts at $t = \tau_E$; and we have introduced the scalar product of two vectors $|A)$ and $|B)$ in the $\mathcal{L}^2[0, T_{\text{int}}]$ space as the following:

$$(A|B) \equiv \int_0^{T_{\text{int}}} A(t) B(t) dt. \tag{11.63}$$

In addition, we have defined filtering functions g_1 and g_2 as

$$g_1(t) \equiv \Theta(t - \tau_E) W(t) \cos \phi_{os}(t), \tag{11.64}$$

$$g_2(t) \equiv \Theta(t - \tau_E) W(t) \sin \phi_{os}(t). \tag{11.65}$$

Since all the data can in principle be digitized and stored in hardware, the weight function $W(t)$ can be realized digitally during data processing. In addition, an overall re-scaling of $g_{1,2}(t) \to C_0 g_{1,2}(t)$, with C_0 a time-independent constant, does not affect the verification performance; also, there are multiple ways of achieving a particular set of $g_{1,2}(t)$, by adjusting the phase $\phi_{os}(t)$ of the local oscillator and the weight function $W(t)$.

In light of Eqs. (11.55)–(11.57), we decompose the weighted quantity $\hat{Y}$ [cf. Eq. (11.62)] as a signal $\hat{Y}_s$ and a noise part $\delta \hat{Y}$, namely:

$$\hat{Y} = \hat{Y}_s + \delta \hat{Y}. \tag{11.66}$$

These are given by

$$\begin{aligned} \hat{Y}_s &= \sqrt{1-\eta}\,(\alpha/\hbar)\,(g_2|\hat{x}_q), \\ \delta \hat{Y} &= (g_1|\delta \hat{b}_1) + (g_2|\delta \hat{b}_2). \end{aligned} \tag{11.67}$$

11.6 State Verification in the Presence of Markovian Noises

Since an overall normalization of $g_{1,2}$ will not affect the signal-to-noise ratio as mentioned, we can mathematically impose that:

$$(g_2|f_1) = \cos\zeta, \quad (g_2|f_2) = \sin\zeta \tag{11.68}$$

with

$$f_1(t) \equiv \cos\omega_m t, \quad f_2(t) \equiv (\Omega_q/\omega_m)\sin\omega_m t \tag{11.69}$$

in the coordinate representation. The signal part can then be rewritten as:

$$\hat{Y}_s = \sqrt{1-\eta}\,(\alpha/\hbar)\delta x_q\left[\hat{x}_0\cos\zeta + \hat{p}_0\sin\zeta\right], \tag{11.70}$$

where we have introduced normalized the oscillator position and momentum as $\hat{x}_0 \equiv \hat{x}(\tau_E)/\delta x_q$ and $\hat{p}_0 \equiv \hat{p}(\tau_E)/\delta p_q$. In such a way, a mechanical quadrature of $\hat{X}_\zeta$ would be probed [cf. Eq. (11.4)]. For the noise part, more explicitly, we have [cf. Eqs. (11.55)–(11.57)]

$$\begin{aligned}\delta\hat{Y} &= (g_1|\sqrt{\eta}\,\hat{n}_1 + \sqrt{1-\eta}\,\hat{a}_1) + (g_2|\sqrt{\eta}\,\hat{n}_2 + \sqrt{1-\eta}\,\hat{a}_2) \\ &+ \sqrt{1-\eta}\,(\alpha^2/\hbar)(g_2|\mathbf{G}_x|\hat{a}_1) \\ &+ \sqrt{1-\eta}\,(\alpha/\hbar)[(g_2|\mathbf{G}_x|\hat{\xi}_F) + (g_2|\hat{\xi}_x)],\end{aligned} \tag{11.71}$$

where the integration with $G_x(t-t')$ has been augmented into applying a linear operator $\mathbf{G}_x$ in the $\mathcal{L}^2[0, T_{\text{int}}]$ space. In the above equation, terms on the first line are the shot noise, the term on the second line is the back-action noise, while terms on the third line are the classical-force and sensing noises.

The optimal $g_1(t)$ and $g_2(t)$ that give a sub-Heisenberg accuracy for each quadrature will be rigorously derived for general situations in the next section. If $\hat{a}_1$ and $\hat{a}_2$ are uncorrelated and there is no readout loss with $\eta = 0$, an optimal choice for g_1 would need to cancel the entire contribution from the back-action noise term (proportional to $\hat{a}_1$). This is equivalent to impose, mathematically, that

$$(g_1|\hat{a}_1) + (\alpha^2/\hbar)(g_2|\mathbf{G}_x|\hat{a}_1) = 0 \tag{11.72}$$

or

$$|g_1) + (\alpha^2/\hbar)\,\mathbf{G}_x^{\text{adj}}|g_2) = 0, \tag{11.73}$$

where $\mathbf{G}_x^{\text{adj}}$ is the adjoint of $\mathbf{G}_x$. Physically, this corresponds to bringing in a piece of shot noise $(g_1|\hat{a}_1)$ to cancel the back-action noise $(\alpha^2/\hbar)(g_2|\mathbf{G}_x|\hat{a}_1)$—therefore achieving a shot-noise-limited only measurement. In the coordinate representation, Eq. (11.73) can be written out more explicitly as:

$$g_1(t) + (\alpha^2/\hbar)\int_t^{T_{\text{int}}} dt'\, G_x(t'-t)g_2(t') = 0, \tag{11.74}$$

which agrees exactly with the variational-type BAE measurement scheme first investigated by Vyatchanin et al. [58]. It is suitable for detecting signals with *known arrival time*. For stationary signals, one would prefer frequency-domain variational techniques proposed by Kimble et al. [32], which evades the back-action noise for all possible signals, as long as they are Gaussian and stationary.

As realized by Kimble et al. [32] in their frequency-domain treatment, when the readout loss is significant (large η) and when the back-action noise is strong (large α), the variational approach becomes less effective, because in such a case, the magnitude of g_1 required to bring enough $\hat{a}_1$ to cancel the back-action noise would also introduce significant noise $\hat{n}_1$ [cf. Eq. (11.71)]. This reasoning apparently leads to a trade-off between the need to evade back action, and the need to minimize loss-induced shot noise—such an optimization will be made in the next section.

11.6.2 Optimal Verification Scheme and Covariance Matrix for the Added Noise: Formal Derivation

Imposing the BAE condition [cf. Eq. (11.74)] does not specify the shape of g_2, nor does Eq. (11.68), and so we have further freedom in choosing the g_2 that minimizes the noise in measuring a particular quadrature of $\hat{X}_\zeta$. In addition, in the presence of readout loss with $\eta \neq 0$, totally evading back action is not the obvious optimum, as mentioned. Therefore, we need to optimize g_1 and g_2 simultaneously. In this section, we first carry out this procedure formally, and then we apply to the Markovian-noise budget in the next subsection.

The total $\hat{x}_q$-referred noise in the weighted estimator $\hat{Y}$ can be written as [cf. Eqs. (11.70) and (11.71)]

$$\sigma^2[g_{1,2}] = \frac{\hbar^2}{(1-\eta)\alpha^2\delta x_q^2} \langle \delta\hat{Y}\delta\hat{Y}\rangle_{\text{sym}}$$

$$= \frac{2}{(1-\eta)\Omega_q} \sum_{i,j=1}^{2} (g_i|\mathbf{C}_{ij}|g_j), \qquad (11.75)$$

where the correlation functions $\mathbf{C}_{ij}$ among the noises are the following:

$$\mathbf{C}_{ij}(t,t') \equiv \langle \delta\hat{b}_i(t)\delta\hat{b}_j(t')\rangle_{\text{sym}}, \quad (i,j=1,2). \qquad (11.76)$$

The optimal $g_{1,2}(t)$ that minimize σ^2 can be obtained through the standard constraint variational method. For this, we define an effective functional as

$$\begin{aligned}\mathcal{J}_{\text{eff}} &= (1-\eta)(\Omega_q/4)\sigma^2[g_{1,2}] - \mu_1(f_1|g_2) - \mu_2(f_2|g_2) \\ &= \frac{1}{2}\sum_{i,j}(g_i|\mathbf{C}_{ij}|g_j) - (\mu_1 f_1 + \mu_2 f_2|g_2),\end{aligned} \qquad (11.77)$$

11.6 State Verification in the Presence of Markovian Noises

where μ_1 and μ_2 are the Lagrange multipliers due to the normalization constraints in Eq. (11.68). Requiring the functional derivative of $\mathcal{J}_{\text{eff}}$ with respect to g_1 and g_2 to be equal to zero, we obtain:

$$\mathbf{C}_{11}|g_1) + \mathbf{C}_{12}|g_2) = 0, \tag{11.78}$$

$$\mathbf{C}_{21}|g_1) + \mathbf{C}_{22}|g_2) = |\mu_1 f_1 + \mu_2 f_2). \tag{11.79}$$

Here $\mathbf{C}_{ij}$ should be viewed as operators in the $\mathcal{L}^2[0, T_{\text{int}}]$ space. This leads to formal solutions for $g_{1,2}$, namely:

$$|g_1) = -\mathbf{C}_{11}^{-1}\mathbf{C}_{12}|g_2), \tag{11.80}$$

$$|g_2) = \mathbf{M}|\mu_1 f_1 + \mu_2 f_2), \tag{11.81}$$

where we have defined

$$\mathbf{M} \equiv \left[\mathbf{C}_{22} - \mathbf{C}_{21}\mathbf{C}_{11}^{-1}\mathbf{C}_{12}\right]^{-1}. \tag{11.82}$$

Re-imposing Eq. (11.68), the unknown Lagrange multipliers $\mu_{1,2}$ can be solved, which are related to ζ by

$$\begin{bmatrix} (f_1|\mathbf{M}|f_1) & (f_1|\mathbf{M}|f_2) \\ (f_2|\mathbf{M}|f_1) & (f_2|\mathbf{M}|f_2) \end{bmatrix} \begin{bmatrix} \mu_1 \\ \mu_2 \end{bmatrix} = \begin{bmatrix} \cos\zeta \\ \sin\zeta \end{bmatrix}. \tag{11.83}$$

Correspondingly, the minimum $\sigma_{\min}^2$ has the following quadratic form:

$$\sigma_{\min}^2 = [\cos\zeta \ \sin\zeta]\mathbf{V}_{\text{norm}}^{\text{add}}\begin{bmatrix} \cos\zeta \\ \sin\zeta \end{bmatrix}. \tag{11.84}$$

Here, the normalized $\mathbf{V}_{\text{norm}}^{\text{add}}$ is a 2×2 covariance matrix, and it is given by:

$$\mathbf{V}_{\text{norm}}^{\text{add}} = \frac{2}{(1-\eta)\Omega_q}\begin{bmatrix} (f_1|\mathbf{M}|f_1) & (f_1|\mathbf{M}|f_2) \\ (f_2|\mathbf{M}|f_1) & (f_2|\mathbf{M}|f_2) \end{bmatrix}^{-1}. \tag{11.85}$$

This relates to the initial definition of the covariance matrix for the added verification noise [cf. Eq. (11.8)] simply by:

$$\mathbf{V}^{\text{add}} = \text{Diag}[\delta x_q, \delta p_q]\mathbf{V}_{\text{norm}}^{\text{add}}\text{Diag}[\delta x_q, \delta p_q]. \tag{11.86}$$

Due to the linearity in Eqs. (11.79) and (11.83), the optimal $g_{1,2}$ for a given quadrature ζ can also be rewritten formally as:

$$g_{1,2}^\zeta = g_{1,2}^X \cos\zeta + g_{1,2}^P \sin\zeta, \tag{11.87}$$

with $g_{1,2}^X \equiv g_{1,2}^\zeta(0)$ and $g_{1,2}^P \equiv g_{1,2}^\zeta(\pi/2)$. Such ζ-dependence of $g_{1,2}$ manifests the fact that a sub-Heisenberg tomography requires different filtering functions, or equivalently different measurement setups, for different quadratures.

11.6.3 Optimal Verification Scheme With Markovian Noise

Given Makovian noises, the corresponding correlation functions for the output noise $\delta\hat{b}_i$ can be written out explicitly as [cf. Eqs. (11.14), (11.19), (11.20), and (11.76)]:

$$C_{11}(t,t') = \frac{\eta + (1-\eta)e^{2q}}{2}\delta(t-t'), \tag{11.88}$$

$$C_{12}(t,t') = C_{21}(t',t) = (1-\eta)\frac{e^{2q}\alpha^2}{2\hbar}G_x(t'-t), \tag{11.89}$$

$$C_{22}(t,t') = \frac{\Lambda^2}{4}\delta(t-t') + (1-\eta)\frac{\alpha^4}{\hbar^2}\left(\frac{e^{2q}}{2}+\zeta_F^2\right)\int_0^\infty dt_1 G_x(t-t_1)G_x(t'-t_1), \tag{11.90}$$

with $\Lambda \equiv \sqrt{2[\eta+(1-\eta)(e^{-2q}+2\zeta_x^2)]}$. Substituting these C_{ij} into Eqs. (11.80) and (11.81), we can obtain the equations for the optimal filtering functions g_1 and g_2. Specifically, for g_1, we have [cf. Eq. (11.80)]

$$g_1(t) + \frac{(1-\eta)e^{2q}}{\eta+(1-\eta)e^{2q}}\frac{\alpha^2}{\hbar}\int_t^{T_{\text{int}}} dt' G_x(t'-t)g_2(t') = 0. \tag{11.91}$$

For g_2, by writing out **M** explicitly, this gives [cf. Eq. (11.81)]

$$\frac{\Lambda^2}{4}g_2(t) + \zeta_F'^2\frac{\alpha^4}{\hbar^2}\iint_0^{T_{\text{int}}} dt' dt_1 G_x(t-t_1)G_x(t'-t_1)g_2(t')$$
$$= \mu_1 f_1(t) + \mu_2 f_2(t), \tag{11.92}$$

where we have introduced ζ_F', which is given by:

$$\zeta_F' \equiv \left[\frac{\eta(1-\eta)e^{2q}}{2[\eta+(1-\eta)e^{2q}]} + (1-\eta)\zeta_F^2\right]^{1/2} \approx \left[\frac{\eta}{2}+\zeta_F^2\right]^{1/2} \tag{11.93}$$

and it is equal to ζ_F for no readout loss. Although here g_1 is still defined in terms of g_2, the optimal verification strategy does not totally evade the back action, as is manifested in the term proportional to η inside the bracket of Eq. (11.93). In the limit of no readout loss with $\eta = 0$, it is identical to the BAE condition in Eq. (11.74). Typically, we have 1% readout loss $\eta = 0.01$, squeezing $e^{2q} = 10$ and $\zeta_F = 0.2$, so this readout loss will only shift ζ_F by 6%, which is negligible. However, if the thermal noise further decreases and/or the measurement strength increases, the effect of readout loss will become significant, entering in a similar way as in the frequency-domain variational measurement proposed by Kimble et al. [32].

The above integral equations for optimal g_1 and g_2 can be solved analytically, as elaborated in Appendix 11.9.3, which in turn gives **M** and the corresponding **V**^{add} [cf. Eqs. (11.80) and (11.85)]. In the free-mass regime with $\Omega_q \gg \omega_m$, closed forms for optimal g_1 and g_2 can be obtained, which, in terms of $g_{1,2}^{X,P}$ [cf. Eq. (11.87)], are given by:

11.6 State Verification in the Presence of Markovian Noises

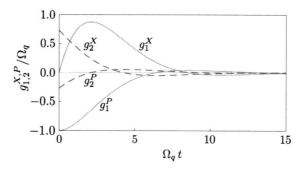

Fig. 11.8 (*color online*) Optimal filtering functions g_1 (*solid curve*) and g_2 (*dashed curve*) in the presence of Markovian noises. We have assumed $\Omega_q/2\pi = 100$ Hz, $\zeta_x = \zeta_F = 0.2$, $\eta = 0.01$ and vacuum input $)q = 0$). For clarity, the origin of the time axis has been shifted from τ_E to 0. From Ref. [39]

$$g_1^X = g_1|_{\zeta=0} = (\Omega_q/\chi)\, e^{-\Omega_q \chi t} \sin \Omega_q \chi t; \tag{11.94}$$

$$g_1^P = g_1|_{\zeta=\frac{\pi}{2}} = -\sqrt{2}\, \Omega_q\, e^{-\Omega_q \chi t} \sin\left(\Omega_q \chi t + \frac{\pi}{4}\right), \tag{11.95}$$

and

$$g_2^X = g_2|_{\zeta=0} = 2\Omega_q \chi\, e^{-\Omega_q \chi t} \cos \Omega_q \chi t; \tag{11.96}$$

$$g_2^P = g_2|_{\zeta=\frac{\pi}{2}} = 2\sqrt{2}\, \Omega_q \chi^2\, e^{-\Omega_q \chi t} \sin\left(\Omega_q \chi t - \frac{\pi}{4}\right), \tag{11.97}$$

with $\chi \equiv [\zeta_F'^2/\Lambda]^{1/2}$. The corresponding verification timescale is set by $\tau_V = (\chi \Omega_q)^{-1}$ and $\tau_q < \tau_V < \tau_F$. To illustrate the behavior of the optimal filtering functions, we show $g_{1,2}^{X,P}$ in Fig. 11.8. As we can see, the verification process finishes after several τ_q, i.e., in a timescale of τ_V.

The corresponding covariance matrix $\mathbf{V}^{\text{add}}$ for the added verification noise is given by

$$\mathbf{V}^{\text{add}} = \frac{1}{1-\eta}\begin{bmatrix} \Lambda^{\frac{3}{2}} \zeta_{F'}^{\frac{1}{2}} \delta x_q^2 & -\Lambda \zeta_F' \hbar/2 \\ -\Lambda \zeta_F' \hbar/2 & 2\Lambda^{\frac{1}{2}} \zeta_{F'}^{\frac{3}{2}} \delta p_q^2 \end{bmatrix}. \tag{11.98}$$

A more summarizing measure of the verification accuracy is the uncertainty product of the added noise ellipse with respect to the Heisenberg limit, namely:

$$U^{\text{add}} = \frac{2}{\hbar}\sqrt{\det \mathbf{V}^{\text{add}}} = \frac{\Lambda \zeta_F'}{1-\eta}. \tag{11.99}$$

In the ideal case with $\eta = 0$, this simply recovers the order-of-magnitude estimate given in Sect. 11.4.4. In Fig. 11.9, we show the uncertainty ellipse for the added noise

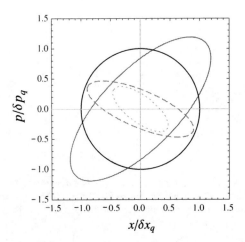

Fig. 11.9 (*color online*) The uncertainty ellipse for the added verification noise in the presence of Markovian noises. We assume $\zeta_x = \zeta_F = 0.2$ with vacuum input (*dashed curve*); and $\zeta_x = \zeta_F = 0.2$ with 10 dB squeezing (*dotted curve*). For contrast, we also show the Heisenberg limit in a unit circle, and the ideal conditional quantum state with a solid ellipse. From Ref. [39]

in the case of $\zeta_x = \zeta_F = 0.2$, readout loss $\eta = 1\%$ and with (green dotted curve) or without (red long-dashed curve) 10 dB input squeezing. In comparison, we also plot the Heisenberg limit (unit circle) and the conditional state obtained through an ideally noiseless state preparation (blue solid ellipse). As figure shows, the least challenging scenario already begins to characterize the conditional quantum state down to the Heisenberg Uncertainty. In these two cases, we have $\Lambda = 1.48$ and 0.62 respectively, leading to:

$$U^{\text{add}} = 0.30 \text{ (vacuum)}, \quad 0.12 \text{ (10 dB squeezing)}. \tag{11.100}$$

11.7 Verification of Macroscopic Quantum Entanglement

In this section, we will apply our protocol to verify macroscopic entanglement between test masses in future GW detectors, which was proposed in Refs. [43, 44]. In the experiment as shown schematically in Fig. 11.10, measurements at the bright and dark port of the interferometer continuously collapse the quantum state of the corresponding common and differential modes of the test-mass motion. This creates two highly squeezed Gaussian states in both modes. Since the common and differential modes are linear combinations of the center of mass motion of test masses in the two arms, namely $\hat{x}^c = \hat{x}^E + \hat{x}^N$ and $\hat{x}^d = \hat{x}^E - \hat{x}^N$, this will naturally generate quantum entanglement between the two test masses, which is similar to creating entanglement by mixing two optical squeezed states at the beam splitter [20, 7].

11.7 Verification of Macroscopic Quantum Entanglement

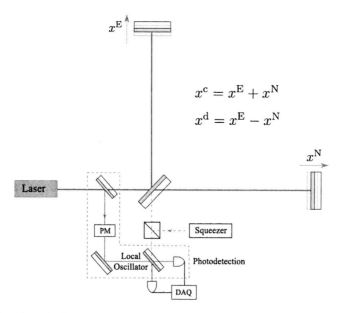

Fig. 11.10 (*color online*) A schematic plot of an advanced interferometric GW detectors for macroscopic entanglement between test masses as a test for gravity decoherence. For simplicity, we have not shown the setup at the bright port, which is identical to that at the dark port. From Ref. [39]

11.7.1 Entanglement Survival Time

To quantify the entanglement strength, we follow Refs. [43, 44] by evaluating the entanglement monotone—the logarithmic negativity defined in Refs. [1, 56]. This can be derived from the covariance matrix for the Gaussian-continuous-variable system considered here. The bipartite covariances among $(\hat{x}^E, \hat{p}^E, \hat{x}^N, \hat{p}^N)$ form the following covariance matrix:

$$\mathbf{V} = \begin{bmatrix} \mathbf{V}_{EE} & \mathbf{V}_{EN} \\ \mathbf{V}_{NE} & \mathbf{V}_{NN} \end{bmatrix}, \tag{11.101}$$

where

$$\mathbf{V}_{EE} = \mathbf{V}_{NN} = \begin{bmatrix} (V_{xx}^c + V_{xx}^d)/4 & (V_{xp}^c + V_{xp}^d)/2 \\ (V_{xp}^c + V_{xp}^d)/2 & (V_{pp}^c + V_{pp}^d) \end{bmatrix}, \tag{11.102}$$

$$\mathbf{V}_{NE} = \mathbf{V}_{EN} = \begin{bmatrix} (V_{xx}^c - V_{xx}^d)/4 & (V_{xp}^c - V_{xp}^d)/2 \\ (V_{xp}^c - V_{xp}^d)/2 & (V_{pp}^c - V_{pp}^d) \end{bmatrix}. \tag{11.103}$$

The logarithmic negativity $E_\mathcal{N}$ can then be written as:

$$E_\mathcal{N} = \max[0, -\log_2 2\sigma_-/\hbar], \tag{11.104}$$

where $\sigma_- \equiv \sqrt{(\Sigma - \sqrt{\Sigma^2 - 4\det \mathbf{V}})/2}$ and $\Sigma \equiv \det \mathbf{V}_{NN} + \det \mathbf{V}_{EE} - 2\det \mathbf{V}_{NE}$. In contrast to Refs. [43, 44], now the covariance matrix $\mathbf{V}$ corresponds to the total covariance matrix $\mathbf{V}^{tot}$ after the entire preparation-evolution-verification process. For Gaussian quantum states, we have [cf. Eqs. (11.10), (11.48) and (11.48)]

$$\mathbf{V}^{tot} = \mathbf{V}(\tau_E) + \mathbf{V}^{add}. \tag{11.105}$$

11.7.2 Entanglement Survival as a Test of Gravity Decoherence

When τ_E increases, the thermal decoherence will increase the uncertainty [cf. Eqs. (11.48) and (11.105)] and eventually the entanglement vanishes, which indicates how long the quantum entanglement can survive. Survival of such quantum entanglement can help us to understand whether there is any additional decoherence effect, such as the *Gravity Decoherence* suggested by Diósi and Penrose [16, 47]. According to their models, quantum superpositions vanish within a timescale of $\hbar/E_G$. Here, E_G can be (a) self-energy of the mass-distribution-difference, namely

$$E_G^{(a)} = \int d\mathbf{x}d\mathbf{y}\, G[\rho(\mathbf{x}) - \rho'(\mathbf{x})][\rho(\mathbf{y}) - \rho'(\mathbf{y})]/r, \tag{11.106}$$

with ρ denoting the mass density distribution and $r \equiv |\mathbf{x} - \mathbf{y}|$; Alternatively, it can be (b) spread of mutual gravitational energy among components of the quantum superposition, namely

$$E_G^{(b)} = \int d\mathbf{x}d\mathbf{y}\, G\rho(\mathbf{x})\rho'(\mathbf{y})\, \delta r/r^{3/2}. \tag{11.107}$$

with δr denoting the uncertainty in location. For the prepared test-mass quantum states with width of δx_q, we have

$$\tau_G^{(a)} \approx \Omega_q/(G\rho), \quad \tau_G^{(b)} \approx \hbar^{1/2}L^2\Omega_q^{1/2}/(Gm^{3/2}). \tag{11.108}$$

where L is the distance between two test masses. Substituting the typical values for LIGO mirrors with $\rho = 2.2\,g/cm^3$, the separation between the two input test masses, $L \approx 10\,m$, and $m = 10\,kg$, we have:

$$\tau_G^a = 4.3 \times 10^9\,s, \quad \tau_G^b = 1.2 \times 10^{-5}\,s. \tag{11.109}$$

It is therefore quite implausible to test model (a); while for model (b), $\Omega_q \tau_G^{(b)}$ is less 0.01 with $\Omega_q/2\pi = 100$Hz. In Fig. 11.11, we show the entanglement survival as a function of evolution duration. As we can see, the model (b) of gravity decoherence can easily be tested, as the entanglement can survive for several times the measurement timescale τ_q, which is much longer than the predicted $\tau_G^{(b)}$.

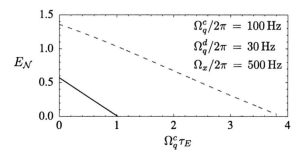

Fig. 11.11 (*color online*) Logarithmic negative $E_\mathcal{N}$ as a function of the evolution duration τ_E, which indicates how long the entanglement survives. The solid curve corresponds to the case where $\Omega_F/2\pi = 20$ Hz and the dashed curve for $\Omega_F/2\pi = 10$ Hz. To maximize the entanglement, the common mode is 10 dB phase squeezed for $t > \tau_E$ and $t < 0$, while the differential mode is 10 dB amplitude squeezed at $t < 0$ and switching to 10 dB phase-squeezed at $t > \tau_E$. From Ref. [39]

11.8 Conclusions

We have investigated in great details a follow-up verification stage after the state preparation and evolution. We have showed the necessity of a sub-Heisenberg verification accuracy in probing the prepared conditional quantum state, and how to achieve it with an optimal time-domain homodyne detection. Including this essential building block—a sub-Heisenberg verification, we are able to outline a complete procedure of a three-staged experiment for testing macroscopic quantum mechanics. In particular, we have been focusing on the relevant free-mass regime and have applied the techniques to discuss MQM experiments with future GW detectors. However, the system dynamics that have been considered describe general cases with a high-Q mechanical oscillator coupled to coherent optical fields. In this respect, we note that our results for Markovian systems only depend on the ratio between various noises and the SQL, and therefore carries over directly to systems on other scales. In addition, the Markovian assumption applies more accurately to smaller-scale systems which operate at higher frequencies.

11.9 Appendix

11.9.1 Necessity of a Sub-Heisenberg Accuracy for Revealing Non-Classicality

As we have mentioned in the introduction, a sub-Heisenberg accuracy is a necessary condition to probe the non-classicality, if the Wigner function of the prepared quantum state has some negative regions, which do not have any classical counterpart.

To prove this necessity, we use the relation between the Q function and the Wigner function as pointed out by Khalili [31]. Given density matrix $\hat{\rho}$, the Q function in the coherent state basis $|\alpha\rangle$ is equal to [23, 53, 59]:

$$Q = \frac{1}{\pi}\langle\alpha|\hat{\rho}|\alpha\rangle, \tag{11.110}$$

which is always positive defined. This is the Fourier transform of the following characteristic function:

$$J(\beta, \beta^*) = \text{Tr}[e^{i\beta^*\hat{a}}e^{i\beta\hat{a}^\dagger}\hat{\rho}]. \tag{11.111}$$

Here, $\hat{a}$ is the annihilation operator and is related to the normalized oscillator position $\hat{x}/\delta x_q$ and momentum $\hat{p}/\delta p_q$ [cf. Eq. (11.26)] by the standard relation:

$$\hat{a} = [(\hat{x}/\delta x_q) + i(\hat{p}/\delta p_q)]/2. \tag{11.112}$$

If we introduce the real and imaginary parts of β, namely, $\beta = \beta_r + i\beta_i$, the characteristic function J can be rewritten as:

$$J(\beta_r, \beta_i) = e^{-(\beta_r^2+\beta_i^2)/2}\text{Tr}[e^{i\beta_r(\hat{x}/\delta x_q)+i\beta_i(\hat{p}/\delta p_q)}\hat{\rho}], \tag{11.113}$$

where we have used the fact that $e^{\hat{A}}e^{\hat{B}} = e^{\hat{A}+\hat{B}}e^{[\hat{A},\hat{B}]/2}$, as $[\hat{A}, \hat{B}]$ commutes with $\hat{A}$ and $\hat{B}$. Inside the bracket of Eq. (11.113), it is the characteristic function for the Wigner function $W(x,p)$, and thus:

$$J(\beta_r, \beta_i) = \frac{1}{(2\pi)^2}\int dx'dp' e^{-(\beta_r^2+\beta_i^2)/2} \\ e^{-i\beta_r(x'/\delta x_q)-i\beta_i(p'/\delta p_q)}W(x', p'). \tag{11.114}$$

Integrating over β_r and β_i, the resulting Q function is given by:

$$Q(x, p) = \frac{1}{2\pi}\int dx'dp' e^{-\frac{1}{2}\left[\frac{(x-x')^2}{\delta x_q^2} + \frac{(p-p')^2}{\delta p_q^2}\right]}W(x', p'). \tag{11.115}$$

This will be the same as Eq. (11.7), if we identify $W_{\text{recon}}(x, p)$ with $Q(x, p)$ and

$$\mathbf{V}^{\text{add}} = \begin{bmatrix} \delta x_q^2 & 0 \\ 0 & \delta p_q^2 \end{bmatrix}, \tag{11.116}$$

which is a Heisenberg-limited error. Since squeezing and a rotation of $\hat{x}$ and $\hat{p}$ axes will not change the positivity of the Q function, Eq. (11.115) basically dictates that the reconstructed Wigner function will always be positive if a Heisenberg-limited error is introduced during the verification stage. Therefore, only if a sub-Heisenberg accuracy is achieved will we be able to reveal the non-classicality of the prepared quantum state.

11.9.2 Wiener-Hopf Method for Solving Integral Equations

In this appendix, we will introduce the mathematical method invented by N. Wiener and E. Hopf for solving a special type of integral equations. For more details, one can refer to a comprehensive presentation of this method and its applications by B. Noble [46]. Here, we will focus on integral equations that can be brought into the following form, as encountered in obtaining the optimal verification scheme:

$$\int_0^{+\infty} dt' C(t, t') g(t') = h(t), \quad t > 0. \tag{11.117}$$

with

$$C(t, t') = A(t - t') + \sum_\alpha \int_0^{\min[t, t']} dt'' B_\alpha^*(t - t'') B_\alpha(t' - t''), \tag{11.118}$$

where $\alpha = 1, 2, \ldots$ and $B_\alpha(t) = 0$ if $t < 0$.

Assuming that solution for $g(t)$ to be a square-integrable function in $\mathcal{L}^2(-\infty, \infty)$, one can split it into *causal* and *anticausal* parts as:

$$g(t) = g_+(t) + g_-(t), \tag{11.119}$$

where $g_-(t)$ is *causal* part:

$$g_-(t) = \begin{cases} 0, t > 0 \\ g(t), t \leqslant 0 \end{cases} \tag{11.120}$$

and $g_+(t)$ is the *anticausal* part of $g(t)$:

$$g_+(t) = \begin{cases} g(t), t > 0 \\ 0, t \leqslant 0. \end{cases} \tag{11.121}$$

This definition enables us to expand the limits of integration in (11.117) and (11.118) to the full range $-\infty < (t, t', t'') < \infty$:

$$\int_{-\infty}^{+\infty} dt' C(t, t') g_+(t') = h(t), \quad t > 0, \tag{11.122}$$

where

$$C(t, t') = A(t - t') + \sum_\alpha \int_{-\infty}^{+\infty} dt'' [B_{\alpha,+}^*(t - t'') B_{\alpha,+}(t' - t'')]_{(+,t'')}, \tag{11.123}$$

and the index $(+, t'')$ stands for taking the causal part of a multidimensional function in the argument t''.

Let us first utilize the method in a simple special case when $B_\alpha(t) \equiv 0, \forall \alpha$, this gives a conventional Wiener-Hopf integral equation:

$$\int_0^{+\infty} dt'\, A(t-t')g(t') = h(t)\,, \quad t > 0, \tag{11.124}$$

which can be rewritten as:

$$\left[\int_{-\infty}^{+\infty} dt'\, A(t-t')g_+(t') - h(t)\right]_{(+,t)} = 0. \tag{11.125}$$

Applying a Fourier transform in t, and the convolution theorem, one gets:

$$\int_{-\infty}^{+\infty} \frac{d\Omega}{2\pi}\left[\tilde{A}(\Omega)\tilde{g}_+(\Omega) - \tilde{h}(\Omega)\right]_+ e^{-i\Omega t} = 0. \tag{11.126}$$

The spectrum of the causal (anticausal) function is simply:

$$\tilde{g}_{+(-)}(\Omega) = \int_{-\infty}^{\infty} dt\, g_{+(-)}(t)e^{i\Omega t}. \tag{11.127}$$

However, this evident relation is not operational for us, as it provides no intuition on how to directly get $\tilde{g}_\pm(\Omega)$ given $\tilde{g}(\Omega)$ at our disposal. The surprisingly simple answer gives complex analysis. Without loss of generality, we can assume that $g(t)$ asymptotically goes to zero at infinity as: $\forall t : |g(t)| < e^{-\gamma_0|t|}$ where γ_0 is some arbitrary positive number, that guarantees regularity of $\tilde{g}(\Omega)$ at $-\infty < \Omega < \infty$. In terms of the analytic continuation $\tilde{g}(s)$ of $\tilde{g}(\Omega)$ to the complex plane $s = \Omega + i\gamma$, the above assumption means that all the poles of $\tilde{g}(s)$ are located outside its band of analyticity $-\gamma_0 < \text{Im}(s) < \gamma_0$. Thus, the partition into causal and anticausal parts for $\tilde{g}(s)$ is now evident:

$$\tilde{g}(s) = \tilde{g}_+(s) + \tilde{g}_-(s) \tag{11.128}$$

where $\tilde{g}_+(s)(\tilde{g}_-(s))$ stands for the function equal to $\tilde{g}(s)$ for $\gamma > \gamma_0(< -\gamma_0)$ and is analytic in the half plane above (below) the line $\gamma = \gamma_0(-\gamma_0)$.[3] According to properties of analytic continuation, this decomposition is unique and completely determined by values of $\tilde{g}(\Omega)$ on the real axis. Moreover, as a Fourier transform of valid $\mathcal{L}^2$-function, it has to approach zero when $|s| \to \infty$. For more general cases, this requirement could be relaxed to demand that ∞ should be a regular point of $\tilde{g}(s)$ so that $\lim_{|s|\to\infty} \tilde{g}(s) = const$. This allows to include δ-function and other integrable distributions into consideration, though it forces us to add the constant $g(\infty)$ to formula (11.128) as additional term. For example, for $g(t) = e^{-\alpha|t|}$, $\alpha > 0$ one has the following Fourier transform:

$$\tilde{g}(s) = \frac{2\alpha}{\alpha^2 + s^2} = \frac{2\alpha}{(s+i\alpha)(s-i\alpha)} \tag{11.129}$$

---

[3] Functions $\tilde{g}_+(s)$ and $\tilde{g}_-(s)$ are, in essence, Laplace transforms of $g(t)$ for positive and negative time, respectively, with only a substitution of the variable $s \to ip$.

11.9 Appendix

that has one pole $s_+ = -i\alpha$ in the lower half complex plane (LHP) and one $s_- = +i\alpha$ in the upper half complex plane (UHP). To split $\tilde{f}(\Omega)$ in accordance with (11.128) one can use the well-known formula:

$$\tilde{g}_\pm(s) = \sum_{\{s_{\pm,k}\}} \frac{\text{Res}[\tilde{g}(s), s_{\pm,k}]}{(s - s_{\pm,k})^{\sigma_k}} \tag{11.130}$$

where summation goes over all poles $\{s_{+,k}\}$ (with σ_k is the order of pole $s_{+,k}$) of $\tilde{g}(s)$ that belong to the LHP for $\tilde{g}_+(s)$ and over all poles $\{s_{-,k}\}$ of $\tilde{g}(s)$ that belong to the UHP for $\tilde{g}_-(s)$ otherwise, and $\text{Res}[\tilde{g}(s), s]$ stands for residue of $\tilde{g}(s)$ at pole s. For our example function this formula gives:

$$\tilde{g}_+(s) = \frac{i}{s + i\alpha}, \quad \tilde{g}_-(s) = -\frac{i}{s - i\alpha}. \tag{11.131}$$

Using the residue theorem, one can easily show that:

$$g_+(t) = e^{-\alpha t}, \text{ for } t > 0 \tag{11.132}$$

$$g_-(t) = e^{\alpha t}, \text{ for } t < 0. \tag{11.133}$$

Coming back to the Eq. (11.126), assume that function $\tilde{A}(\Omega)$ can be factorized in the following way:

$$\tilde{A}(\Omega) = \tilde{a}_-(\Omega)\tilde{a}_+(\Omega) \tag{11.134}$$

where $\tilde{a}_{+(-)}(\Omega)$ is a function analytic in the UHP (LHP) with its inverse, i.e., both its poles and zeroes are located in the LHP (UHP). One gets the following equation:

$$\left[\tilde{a}_-(\Omega)\tilde{a}_+(\Omega)\tilde{g}_+(\Omega) - \tilde{h}(\Omega)\right]_+ = 0. \tag{11.135}$$

To solve this equation, one realizes the following fact: for any function $\tilde{f}$, $[\tilde{f}(\Omega)]_+ = 0$ means that $\tilde{f}$ has no poles in the LHP. Multiplication of $\tilde{f}$ by any function $\tilde{g}_-$ which also has no poles in the LHP will evidently not change the equality, namely, $[\tilde{g}_-(\Omega)\tilde{f}(\Omega)]_+ = 0$. Multiplying Eq. (11.135) by $1/\tilde{a}_-(\Omega)$, the solution reads

$$\tilde{g}_+(\Omega) = \frac{1}{\tilde{a}_+(\Omega)} \left[\frac{\tilde{h}(\Omega)}{\tilde{a}_-(\Omega)}\right]_+. \tag{11.136}$$

On performing an inverse Fourier transform of $\tilde{g}_+(\Omega)$, the time-domain solution $g_+(t)$ can be obtained.

Now we are ready to solve Eq. (11.22) with the general kernel in Eq. (11.123). Performing similar manipulations, one obtains the following equation for $\tilde{g}_+(\Omega)$ in the Fourier domain:

$$\left[\left(\tilde{A} + \sum_\alpha \tilde{B}_\alpha \tilde{B}_\alpha^*\right)\tilde{g}_+ - \sum_\alpha \tilde{B}_\alpha (\tilde{B}_\alpha^* \tilde{g}_+)_- - \tilde{h}\right]_+ = 0, \quad (11.137)$$

where we have omitted arguments Ω of all functions for brevity. Since $\tilde{B}_\alpha$ is a causal function, $\tilde{B}_\alpha^*$ is anticausal and $\tilde{g}_+$ is causal, $()\tilde{B}_\alpha^* \tilde{g})_-$ only depends on the value of $\tilde{g}$ on the poles of $\tilde{B}_\alpha^*$. Performing a similar factorization:

$$\tilde{\psi}_+ \tilde{\psi}_- = \tilde{A} + \sum_\alpha \tilde{B}_\alpha \tilde{B}_\alpha^*, \quad (11.138)$$

with $\tilde{\psi}_+$ ($\tilde{\psi}_-$) and $1/\tilde{\psi}_+$ ($1/\tilde{\psi}_-$) analytic in the UHP (LHP), $\psi_+(-\Omega) = \psi_+^*(\Omega) = \psi_-(\Omega)$, we get the solution in the form:

$$\tilde{g}_+ = \frac{1}{\tilde{\psi}_+}\left[\frac{\tilde{h}}{\tilde{\psi}_-}\right]_+ + \frac{1}{\tilde{\psi}_+}\left[\sum_\alpha \frac{\tilde{B}_\alpha(\tilde{B}_\alpha^* \tilde{g}_+)_-}{\tilde{\psi}_-}\right]_+. \quad (11.139)$$

Even though $\tilde{g}_+$ also enters the right hand side of the above equation, yet $(\tilde{B}_\alpha^* \tilde{g}_+)_-$ can still be written out explicitly as:

$$(\tilde{B}_\alpha^* \tilde{g}_+)_- = \sum_{\{\Omega_{-,k}\}} \frac{\tilde{g}_+(\Omega_{-,k})\mathrm{Res}[\tilde{B}^*(\Omega), \Omega_{-,k}]}{(\Omega - \Omega_{-,k})^{\sigma_k}}. \quad (11.140)$$

Here $\{\Omega_{-,k}\}$ are poles of $\tilde{B}^*(\Omega)$ that belong to UHP, and therefore $\tilde{g}_+(\Omega_{-,k})$ are just constants that can be obtained by solving a set of linear algebra equations, by evaluating Eq. (11.139) at those poles $\{\Omega_{-,k}\}$.

11.9.3 Solving Integral Equations in Section 11.6

Here, we will use the technique introduced in the previous section to obtain analytical solutions to the integral equations we encountered in Sects. 11.6.2 and 11.6.3.

In the coordinate representation, the integral equations for $g_{1,2}$ are the following [cf. Eqs. (11.78) and (11.79)]:

$$\int_0^{T_\mathrm{int}} dt' \begin{bmatrix} C_{11}(t,t') & C_{12}(t,t') \\ C_{21}(t,t') & C_{22}(t,t') \end{bmatrix} \begin{bmatrix} g_1(t') \\ g_2(t') \end{bmatrix} = \begin{bmatrix} 0 \\ h(t) \end{bmatrix}, \quad (11.141)$$

where C_{ij} ($i, j = 1, 2$) are given by Eqs. (11.88), (11.89) and (11.90), and we have defined $h(t) \equiv \mu_1 f_1(t) + \mu_2 f_2(t)$. Since the optimal $g_{1,2}(t)$ will automatically cut off when $t > \tau_F$, we can extend the integration upper bound T_int to ∞. This brings the equations into the appropriate form considered in Appendix 11.9.2. In the frequency domain, they can be written as

11.9 Appendix

$$[\tilde{S}_{11}\tilde{g}_1]_+ + [\tilde{S}_{12}\tilde{g}_2]_+ = 0, \tag{11.142}$$

$$[\tilde{S}_{21}\tilde{g}_1]_+ + [\tilde{S}_{22}\tilde{g}_2]_+ - \tilde{\Gamma} = \tilde{h}, \tag{11.143}$$

$$\tilde{\Gamma} = (1-\eta)(\Omega_q^4/2)(e^{2q} + 2\zeta_F^2)[\tilde{G}_x(\tilde{G}_x\tilde{g}_2)_-]_+. \tag{11.144}$$

Here, $\tilde{S}_{ij}$ are the Fourier transformations of the correlation functions $\mathbf{C}_{ij}$. Specifically, they are

$$\tilde{S}_{11} = \frac{\eta + (1-\eta)e^{2q}}{2}, \tag{11.145}$$

$$\tilde{S}_{12} = -\frac{(1-\eta)e^{2q}\Omega_q^2}{2(\Omega + \omega_m - i\gamma_m)(\Omega - \omega_m - i\gamma_m)}, \tag{11.146}$$

$$\tilde{S}_{21} = \tilde{S}_{12}^*, \tag{11.147}$$

$$\tilde{S}_{22} = \frac{\Lambda^2}{4} + \frac{(1-\eta)(e^{2q} + 2\zeta_F^2)\Omega_q^4}{2[(\Omega + \omega_m)^2 + \gamma_m^2][(\Omega - \omega_m)^2 + \gamma_m^2]}. \tag{11.148}$$

Since $\tilde{S}_{11}$ is only a number, the solution to $\tilde{g}_1$ is simply

$$\tilde{g}_1 = -\tilde{S}_{11}^{-1}[\tilde{S}_{12}\tilde{g}_2]_+. \tag{11.149}$$

In the time-domain, this recovers the result in Eq. (11.91). Through a spectral factorization

$$\tilde{\psi}_+\tilde{\psi}_- \equiv \tilde{S}_{22} - \tilde{S}_{11}^{-1}\tilde{S}_{12}\tilde{S}_{21}, \tag{11.150}$$

we obtain the solution for $\tilde{g}_2$:

$$\tilde{g}_2 = \frac{1}{\tilde{\psi}_+}\left\{\frac{1}{\tilde{\psi}_-}\left[\tilde{h} - \tilde{S}_{11}^{-1}\tilde{S}_{21}(\tilde{S}_{12}\tilde{g}_2)_- + \tilde{\Gamma}\right]\right\}_+. \tag{11.151}$$

Substituting $\tilde{\Gamma}$ into the above equation, $\tilde{g}_2$ becomes:

$$\tilde{g}_2 = \frac{1}{\tilde{\psi}_+}\left\{\frac{1}{\tilde{\psi}_-}\left[\tilde{h} + \kappa\,\tilde{G}_x(\tilde{G}_x^*\tilde{g}_2)_-\right]\right\}_+ \tag{11.152}$$

with $\kappa \equiv m^2\Omega_q^4\zeta_F'^2$. A simple inverse Fourier transformation gives $g_1(t)$ and $g_2(t)$. The unknown Lagrange multipliers can be solved using Eq. (11.83). We can then derive the covariance matrix $\mathbf{V}^{\text{add}}$ for the added verification noise with Eq. (11.85). In the free-mass regime, a closed form for $\mathbf{V}^{\text{add}}$ can be obtained, as shown explicitly in Eq. (11.98).

References

1. G. Adesso, F. Illuminati, Entanglement in continuous-variable systems: recent advances and current perspectives. J. Phys. A: Math. Theor. **40**(28), 7821 (2007)
2. O. Arcizet, P.-F. Cohadon, T. Briant, M. Pinard, A. Heidmann, Radiation-pressure cooling and optomechanical instability of a micromirror. Nature **444**, 71–74 (2006)
3. A. Barchielli, Stochastic differential equations anda posteriori states in quantum mechanics. Int. J. Theor. Phys. **32**, 2221–2233 (1993)
4. D.G. Blair, E.N. Ivanov, M.E. Tobar, P.J. Turner, van F. Kann, I.S. Heng, High sensitivity gravitational wave antenna with parametric transducer readout. Phys. Rev. Lett. **74**, 1908 (1995)
5. K.J. Blow, R. Loudon, S.J.D. Phoenix, T.J. Shepherd, Fields in quantum optics. Phys. Rev. A **42**, 4102 (1990)
6. S. Bose, K. Jacobs, P.L. Knight, Preparation of nonclassical states in cavities with a moving mirror. Phys. Rev. A **56**, 4175 (1997)
7. W.P. Bowen, R. Schnabel, P.K. Lam, T.C. Ralph, Experimental investigation of criteria for continuous variable entanglement. Phys. Rev. Lett. **90**, 043601 (2003)
8. V.B. Braginsky, Classical and quantum restrictions on the detection of weak disturbances of a macroscopic oscillator. JETP **26**, 831 (1968)
9. V.B. Braginsky, F.Y. Khalili, *Quantum measurement* (Cambridge University Press, Cambridge, 1992)
10. C.M. Caves, B.L. Schumaker, New formalism for two-photon quantum optics I - Quadrature phases and squeezed states. II - Mathematical foundation and compact notation. Phys. Rev. A **31**, 3068 (1985)
11. P.F. Cohadon, A. Heidmann, M. Pinard, Cooling of a mirror by radiation pressure. Phys. Rev. Lett. **83**, 3174 (1999)
12. LIGO Scientific Collaboration, Observation of a kilogram-scale oscillator near its quantum ground state. New J. Phys. **11**(7), 073032 (2009)
13. T. Corbitt, Y. Chen, E. Innerhofer, H. Mueller-Ebhardt, D. Ottaway, H. Rehbein, D. Sigg, S. Whitcomb, C. Wipf, N. Mavalvala, All-optical trap for a gram-scale mirror. Phys. Rev. Lett. **98**, 150802–150804 (2007)
14. T. Corbitt, C. Wipf, T. Bodiya, D. Ottaway, D. Sigg, N. Smith, S. Whitcomb, N. Mavalvala, Optical dilution and feedback cooling of a gram-scale oscillator to 6.9 mK. Phys. Rev. Lett. **99**, 160801–160804 (2007)
15. S. Danilishin, H. Mueller-Ebhardt, H. Rehbein, K. Somiya, R. Schnabel, K. Danzmann, T. Corbitt, C. Wipf, N. Mavalvala, Y. Chen, Creation of a quantum oscillator by classical control. arXiv:0809.2024, (2008)
16. L. Diosi, Models for universal reduction of macroscopic quantum fluctuations. Phys. Rev. A **40**, 1165 (1989)
17. L. Diosi, Laser linewidth hazard in optomechanical cooling. Phys. Rev. A **78**, 021801 (2008)
18. A.C. Doherty, S.M. Tan, A.S. Parkins, D.F. Walls, State determination in continuous measurement. Phys. Rev. A **60**, 2380 (1999)
19. I. Favero, C. Metzger, S. Camerer, D. Konig, H. Lorenz, J.P. Kotthaus, K. Karrai, Optical cooling of a micromirror of wavelength size. Appl. Phys. Lett. **90**, 104101–104103 (2007)
20. A. Furusawa, J.L. S03rensen, S.L. Braunstein, C.A. Fuchs, H.J. Kimble, E.S. Polzik, Unconditional quantum teleportation. Science **282**, 706–709 (1998)
21. C. Gardiner, P. Zoller, *Quantum noise* (Springer-Verlag, Berlin, 2004)
22. C. Genes, D. Vitali, P. Tombesi, Simultaneous cooling and entanglement of mechanical modes of a micromirror in an optical cavity. New J. Phys. **10**(9), 095009 (2008)
23. S. Gigan, H.R. Böhm, M. Paternostro, F. Blaser, G. Langer, J.B. Hertzberg, K.C. Schwab, D. Bäuerle, M. Aspelmeyer, A. Zeilinger, Self-cooling of a micromirror by radiation pressure. Nature **444**, 67–70 (2006)

References 201

24. S. Gröblacher, S. Gigan, H.R. Böhm, A. Zeilinger, M. Aspelmeyer, Radiation-pressure self-cooling of a micromirror in a cryogenic environment. EPL (Europhys. Lett.) **81**(5), 54003 (2008)
25. S. Gröblacher, J.B. Hertzberg, M.R. Vanner, G.D. Cole, S. Gigan, K.C. Schwab, M. Aspelmeyer, Demonstration of an ultracold microoptomechanical oscillator in a cryogenic cavity. Nat. Phys. **5**, 485–488 (2009)
26. A. Hopkins, K. Jacobs, S. Habib, K. Schwab, Feedback cooling of a nanomechanical resonator. Phys. Rev. B **68**, 235328 (2003)
27. http://www.ligo.caltech.edu.
28. K. Jacobs, P. Lougovski, M. Blencowe, Continuous measurement of the energy eigenstates of a nanomechanical resonator without a nondemolition probe. Phys. Rev. Lett. **98**, 147201 (2007)
29. G. Jourdan, F. Comin, J. Chevrier, Mechanical mode dependence of bolometric backaction in an atomic force microscopy microlever. Phys. Rev. Lett. **101**, 133904 (2008)
30. I. Katz, A. Retzker, R. Straub, R. Lifshitz, Signatures for a classical to quantum transition of a driven nonlinear nanomechanical resonator. Phys. Rev. Lett. **99**, 040404 (2007)
31. F. Khalili, S. Danilishin, H. Miao, H. Mueller-Ebhardt, H. Yang, Y. Chen, Preparing a mechanical oscillator in non-Gaussian quantum states. Submitted to PRL (2010)
32. H.J. Kimble, Y. Levin, A.B. Matsko, K.S. Thorne, S.P. Vyatchanin, Conversion of conventional gravitational-wave interferometers into quantum nondemolition interferometers by modifying their input and/or output optics. Phys. Rev.D **65**, 022002 (2001)
33. D. Kleckner, D. Bouwmeester, Sub-kelvin optical cooling of a micromechanical resonator. Nature **444**, 75–78 (2006)
34. A.I. Lvovsky, M.G. Raymer, Continuous-variable optical quantumstate tomography. Rev. Mod. Phys. **81**, 299 (2009)
35. S. Mancini, V.I. Man'ko, P. Tombesi, Ponderomotive control of quantum macroscopic coherence. Phys. Rev. A **55**, 3042 (1997) Apr
36. S. Mancini, D. Vitali, P. Tombesi, Optomechanical cooling of a macroscopic oscillator by homodyne feedback. Phys. Rev. Lett. **80**, 688 (1998)
37. F. Marquardt, J.P. Chen, A.A. Clerk, S.M. Girvin, Quantum theory of cavity-assisted sideband cooling of mechanical motion. Phys. Rev. Lett. **99**, 093902 (2007)
38. C.H. Metzger, K. Karrai, Cavity cooling of a microlever. Nature **432**, 1002–1005 (2004)
39. H. Miao, S. Danilishin, H. Mueller-Ebhardt, H. Rehbein, K. Somiya, Y. Chen, Probing macroscopic quantum states with a sub-Heisenberg accuracy.. Phys. Rev. A **81**, 012114 (2010)
40. G.J. Milburn, Classical and quantum conditional statistical dynamics. Quantum and Semiclassical Opt.: J. Eur. Opt. Soc. Part B **8**(1), 269 (1996)
41. S. Miyoki, T. Uchiyama, K. Yamamoto, H. Hayakawa, K. Kasahara, H. Ishitsuka, M. Ohashi, K. Kuroda, D. Tatsumi, S. Telada, M. Ando, T. Tomaru, T. Suzuki, N. Sato, T. Haruyama, Y. Higashi, Y. Saito, A. Yamamoto, T. Shintomi, A. Araya, S. Takemoto, T. Higashi, H. Momose, J. Akamatsu, W. Morii, Status of the CLIO project. Classical Quantum Gravity **21**(5), S1173 (2004)
42. C.M. Mow-Lowry, A.J. Mullavey, S. Gossler, M.B. Gray, D.E. Mc- Clelland, Cooling of a gram-scale cantilever flexure to 70 mK with a servo-modified optical spring. Phys. Rev. Lett. **100**, 010801–010804 (2008)
43. H. Mueller-Ebhardt, H. Rehbein, R. Schnabel, K. Danzmann, Y. Chen, Entanglement of macroscopic test masses and the standard quantum limit in laser interferometry. Phys. Rev. Lett. **100**, 013601 (2008)
44. H. Mueller-Ebhardt, H. Rehbein, C. Li, Y. Mino, K. Somiya, R. Schnabel, K. Danzmann, Y. Chen, Quantum-state preparation and macroscopic entanglement in gravitational-wave detectors. Phys. Rev. A **80**, 043802 (2009)
45. A. Naik, O. Buu, M.D. LaHaye, A.D. Armour, A.A. Clerk, M.P. Blencowe, K.C. Schwab, Cooling a nanomechanical resonator with quantum back-action. Nature **443**, 193–196 (2006)
46. B. Noble, Methods based on the Wiener-Hopf technique for the solution of partial differential equations. (AMS Chelsea Publishing, New York, 1988)

47. R. Penrose, On gravity's role in quantum state reduction. Gen. Relativ. Gravitation **28**, 581–600 (1996)
48. M. Poggio, C.L. Degen, H.J. Mamin, D. Rugar, Feedback cooling of a Cantilever's fundamental mode below 5 mK. Phys. Rev. Lett. **99**, 017201–017204 (2007)
49. P. Rabl, C. Genes, K. Hammerer, M. Aspelmeyer, Phase-noise induced limitations on cooling and coherent evolution in optomechanical systems. Phys. Rev. A **80**, 063819 (2009)
50. S.W. Schediwy, C. Zhao, L. Ju, D.G. Blair, P. Willems, Observation of enhanced optical spring damping in a macroscopic mechanical resonator and application for parametric instability control in advanced gravitationalwave detectors. Phys. Rev. A **77**, 013813–013815 (2008)
51. A. Schliesser, P. Del'Haye, N. Nooshi, K.J. Vahala, T.J. Kippenberg, Radiation pressure cooling of a micromechanical oscillator using dynamical backaction. Phys. Rev. Lett. **97**, 243905-4 (2006)
52. A. Schliesser, R. Riviere, G. Anetsberger, O. Arcizet, T.J. Kippenberg, Resolved-sideband cooling of a micromechanical oscillator. Nat. Phys. **4**, 415–419 (2008)
53. M.O. Scully, M.S. Zubairy, *Quantum optics* (Cambridge University Press, Cambridge, 1997)
54. J.D. Teufel, J.W. Harlow, C.A. Regal, K.W. Lehnert, Dynamical backaction of microwave fields on a nanomechanical oscillator. Phys. Rev. Lett. **101**, 197203–197204 (2008) Nov
55. J.D. Thompson, B.M. Zwickl, A.M. Jayich, F. Marquardt, S.M. Girvin, J.G.E. Harris, Strong dispersive coupling of a high-finesse cavity to a micromechanical membrane. Nature **452**, 72–75 (2008)
56. G. Vidal, R.F. Werner, Computable measure of entanglement. Phys. Rev. A **65**, 032314 (2002)
57. S. Vyatchanin, Effective cooling of quantum system. Dokl. Akad. Nauk. SSSR **234**, 688 (1977)
58. S.P. Vyatchanin, E.A. Zubova, Quantum variation measurement of a force. Phys. Lett. A **201**, 269–274 (1995)
59. D.F. Walls, G. Milburn, *Quantum optics* (Springer-Verlag, Berlin, 2008)
60. I. Wilson-Rae, N. Nooshi, W. Zwerger, T.J. Kippenberg, Theory of ground state cooling of a mechanical oscillator using dynamical backaction. Phys. Rev. Lett. **99**, 093901 (2007)
61. B. Yurke, D. Stoler, Measurement of amplitude probability distributions for photon-number-operator eigenstates. Phys. Rev. A **36**, 1955 (1987)
62. C. Zhao, L. Ju, H. Miao, S. Gras, Y. Fan, D.G. Blair, Three-mode optoacoustic parametric amplifier: a tool for macroscopic quantum experiments. Phys. Rev. Lett. **102**, 243902 (2009)

Chapter 12
Conclusions and Future Work

12.1 Conclusions

To conclude, this thesis has covered two main topics concerning Macroscopic Quantum Mechanics (MQM) in optomechanical devices. The first topic considers different approaches to surpassing the Standard Quantum Limit (SQL) for measuring weak forces, which include modifying the input/output optics, and the dynamics of the mechanical oscillator. Concerning the approach of modifying the input optics, in Chap. 3, we have proposed simultaneously injecting two squeezed light—filtered by a resonant optical cavity—into the dark port of the laser interferometer GW detector. This can reduce the low-frequency radiation-pressure noise, and the high-frequency shot noise so that the detector sensitivity over the entire observational band from 10 Hz to 10^4 Hz can be improved. Given its relatively simple setup—only a 30 m filter cavity is required—this could be a feasible add-on to future advanced GW detectors. With the approach of modifying the output optics, in Chap. 5, we have proposed the use of a time-domain variational method, in which the homodyne detection angle has the optimal time-dependence. Such a scheme provides a transparent way to probe the mechanical quadrature—the conserved dynamical quantity of a mechanical oscillator—and allows us to surpass the SQL. This works in the cases where the bandwidth of the optical cavity is much larger than the mechanical frequency, and therefore it can be implemented in small-scale optomechanical devices, in which high finesse is difficult to achieve, and also in large-scale broadband GW detectors (e.g., Advanced LIGO). With the approach of modifying the mechanical dynamics, in Chap. 4, we have explored the frequency dependence in double optical springs, and we have shown that it can significantly enhance the mechanical response over a broad frequency band. This is especially useful for GW detectors because the natural frequency of the suspended test-masses is quite low, which gives a low mechanical response in the detection band around 100 Hz.

The second topic is concerned with exploring quantum behaviors of macroscopic mechanical oscillators with quantum-limited optomechanical devices. In Chap. 6, we have discussed the use of three-mode optomechanical interactions to study MQM.

We have pointed out the optimal frequency matching inherent in this interaction, which allows a significant enhancement of the optomechanical coupling compared with that in conventional two-mode interactions. Such a feature enables us to cool milligram-scale mechanical oscillators down to their quantum ground state and also to create quantum entanglement between the oscillator and the cavity modes. In Chap. 7, we have discussed the quantum limit for ground state cooling, and the creation of quantum entanglement in general optomechanical devices. We have used an alternative point of view, based upon information loss, to explain the origin of the resolved-sideband cooling limit. By recovering the information contained in the cavity output, we can surpass such a cooling limit without imposing stringent requirements on the cavity bandwidth. We have also shown such an information recovery can enhance the optomechanical entanglement. This work can help us find the proper parameter regime to achieve the quantum ground state, and to realize quantum entanglement experimentally. In Chap. 8, we have investigated the quantum entanglement between a mechanical oscillator and a continuum optical field. In contrast to the cases studied in previous chapters, the continuum optical field contains infinite degrees of freedom. We have developed a new functional method to analyze the entanglement strength, and have derived an elegant scaling which shows that the entanglement only depends on ratio of the optomechanical coupling to the thermal decoherence strength. This illuminates the possibility of incorporating mechanical degrees of freedom for future quantum computing at high environmental temperature. In Chap. 9, we have studied nonlinear optomechanical interactions for observing mechanical energy quantization. We have derived a simple quantum limit that only involves the fundamental parameters of an optomechanical device—this requires the zero-point uncertainty of the mechanical oscillator to be comparable to the linear dynamical range of the cavity, which is quantified by the ratio of the optical wavelength to the cavity finesse. This limit applies universally to all optomechanical devices, and therefore it serves as a guiding tool for choosing the right parameters for MQM experiments. In Chap. 10, we have discussed preparing a mechanical oscillator in non-Gaussian quantum states to explore the non-classical features of optomechanical devices. We have proposed transferring a non-Gaussian quantum state from the optical field to the mechanical oscillator by injecting a single-photon pulse into the dark port of an interferometric optomechanical device. The radiation pressure of the single-photon pulse is coherently amplified by the strong optical power from the bright port, which makes such a state transfer possible. We have shown the experimental feasibility in the case of both small-scale table-top experiments, and large-scale advanced GW detectors. In Chap. 11, we have outlined a complete procedure for an MQM experiment, by including a verification stage to follow up the preparation stage. With an optimal time-dependent homodyne detection, the prepared quantum state can be probed and verified with a sub-Heisenberg accuracy. This not only allows us to explore the quantum dynamics of the mechanical oscillator, but also the non-classical feature of the quantum state. In particular, we have applied it to study the survival duration of the quantum entanglement between macroscopic test-masses which suffer from thermal decoherence. This complete procedure can also be directly applied to small-scale optomechanical devices.

12.2 Future Work

There are still many issues that need to be further investigated for a thorough understanding. Particularly, in Chap. 4, only preliminary results are obtained: they only show the modified mechanical response due to the double optical spring but not the resulting detector sensitivity. We need to combine the outputs of the two optical fields in an optimal way to maximize the detector sensitivity. In addition, the resulting system is dynamically unstable and we need to figure out a way to control such an instability. In Chap. 5, we have assumed no optical loss in the variational readout, which certainly fails actually. If the optical loss is included, total evasion of the measurement back-action is not obviously the optimum as we have found when studying the verification problem in Chap. 11. Since the mathematical structure of this problem is quite different from that in the verification case, we need to find a new method to realize such an optimization. In Chap. 9, we have studied the optomechanical dynamics by using the Langevin equation approach. This shows how the dynamical quantities evolve, but the effect of measurement on the evolution of the quantum state is not discussed. Therefore, it is necessary for us to use a different approach to study how the mechanical oscillator jumps among different energy eigenstates, while at the same time being, subjected to both a continuous measurement and thermal decoherence. This can help us to address the question of whether the quantum-Zeno effect exists or not in such a nonlinear optomechanical system. In Chap. 11, we have assumed that the thermal force noise and measurement sensing noise are white with flat noise spectra—a Markovian assumption. Non-Markovian noises certainly arise in an actual GW detector. These noises at low frequencies, such as the suspension thermal noise and the coating thermal noise, tend to rise faster than what we have assumed. We have already developed the right tools to treat such non-Markovianity but need further analysis to go through the details.

CPSIA information can be obtained at www.ICGtesting.com
Printed in the USA
LVOW10s1528020314

375722LV00004B/102/P